A LEVEL
Questions and Answers

CHEMISTRY

Bob McDuell

Senior Examiner

Letts
EDUCATIONAL

SERIES EDITOR: BOB McDUELL

Contents

As a rough guide, in an examination you should be aiming to score a mark each minute. When about 15 minutes of the examination remain, it is worth checking whether you are short of time. If you feel you are seriously running out of time, it is very important to try to score as many marks as possible in the short time that remains. In any question worth five marks, one or two will be easily scored and one or two will be very difficult to score. Concentrate on scoring the easy marks on each question that remains. Do not try to write sentences. Just put the main points down clearly in note form.

DIFFERENT TYPES OF EXAMINATION QUESTION

There are different types of examination question which appear on examination papers. Check in a copy of the syllabus, or with your teacher, to find the types of question which will be used on your syllabus.

Multiple-choice questions

When answering multiple-choice questions you should remember the following points.

❶ There is only one correct answer. Do not give more than one answer to any question.

❷ There is no penalty if you get a question wrong. If you do not know the answer, and cannot work it out, have a guess. If you can rule out some answers as definitely wrong you increase your chances of guessing correctly. Guessing is a legitimate approach to multiple-choice questions.

❸ Read the stem carefully several times. Often a mistake is caused by 'jumping in' without reading the stem completely and giving the answer to the question you think the examiner is asking.

❹ First, try and pick the correct answer in your mind.

❺ Check each of the other answers to make sure they are wrong.

❻ If no other answer seems right, you have your answer to record.

❼ If more than one answer seems to be correct, try to eliminate until you are left with one. If you cannot, it is the stage to guess. Choosing one from two is much better than one from four or five.

❽ Very often your first answer is the right answer. You should consider carefully before changing your mind.

A multiple-choice test is designed to cover a large part of the syllabus. Other types of question may not give such good coverage. You cannot hide the parts of the syllabus you are weak in with multiple-choice tests.

Grid questions

These are another type of multiple-choice question used on Chemistry papers by the Scottish Examination Board.

In these questions you have to circle the appropriate letter or letters in the answer grid, e.g.

A CH_4	B H_2	C CO_2
D CO	E C_2H_5OH	F C

(a) Identify the hydrocarbon(s).

[*The question does not tell you whether one or more than one substance is required. The only correct answer is A and you should circle this letter in the grid. A hydrocarbon is a compound of hydrogen and carbon only. If you circle too many answers you lose marks.*]

3

(b) Identify the substance(s) which can burn to produce **both** carbon dioxide and water.

[*Again the question does not tell you how many answers to give. The correct answers are **A** and **E** – the only ones where carbon and hydrogen are combined together.*]

Structured questions

These are the most common type of question used on A-level Chemistry papers. The reason why they are used so widely is that they are so versatile. They can be short, with little opportunity for extended writing. Alternatively, they can be longer and more complex in their structure, with opportunities for extended writing and the demonstration of higher level skills of interpretation and evaluation.

In a structured question, the question is divided into parts (a), (b), (c), etc. These parts can be further subdivided into (i), (ii), (iii), (iv), etc. A structure is built into the question and hence into your answer. This is where the term 'structured question' comes from.

For each part of the question there are often a number of lines or a space for your answer. This is a guide to you about the detail required in the answer, but it does not have to limit you. If you require more space, continue your answer on a separate sheet of paper, but make sure you label the answer clearly, e.g. 3(a)(ii).

To give you a guide as you work through structured questions, papers are often designed to enable you to score one mark per minute. A question worth a maximum of 15 marks should therefore take about 15 minutes to answer.

You do not have to write your answers in full sentences. Concise notes are often the most suitable response.

Like all questions, it is most important to read the stimulus material in the question thoroughly and more than once. This information is often not used fully by students and, as a result, the question is not answered fully. The key to answering many of these questions comes from an appreciation of the full meaning of the 'command word' at the start of the question – 'state', 'describe', 'explain'. The following glossary of command words may help you in the answering of structured questions.

- **State** This means a brief answer is required, with no supporting evidence. Alternatives include **write down**, **give**, **list**, **name**.
- **Define** Just a definition is required.
- **State and explain** A short answer is required (see **state**), but then an explanation is required. A question of this type should be worth more than one mark.
- **Describe** This is often used with reference to a particular experiment. The important points should be given about each stage. Again this type of question is worth more than one mark.
- **Compare and contrast** In this type of question you need to look at similarities and differences. For example, compare and contrast the reactions of ethene and benzene. Before starting to answer this type of question, do a rough table with headings *similarities* and *differences*.
- **Outline** The answer should be brief and the main points picked out.
- **Predict** A brief answer is required, without supporting evidence. You are required to make logical links between various pieces of information.
- **Complete** You are required to add information to a diagram, sentence, flow chart, graph, key, table, etc.
- **Find** This is a general term which may mean calculate, measure, determine, etc.
- **Calculate** A numerical answer is required. You should show your working in order to get an answer. Do not forget the correct units.
- **Suggest** There is not just one correct answer, or you may be applying your answer to a situation outside the syllabus.

Free response questions

These can include essay questions. In this type of question you are given a question which enables you to develop your answer in different ways. Your answer really is a free response and you write as much as you wish. Candidates often do not write enough or try to 'pad out' the answer.

Remember you can only score marks when your answer matches the marking points on the examiner's marking scheme.

In this type of question it is important to plan your answer before starting it, allocating the correct amount of time to each part of the question.

Comprehension questions

Several Examination Boards set compulsory comprehension questions. These questions often provide 2–3 closely printed A4 pages of information. The answers to many of the questions are given in the passage and other answers can be worked out by understanding the passage. It is most important to read the passage through twice before attempting the questions. It is obvious many candidates go straight into answering the questions. The first time you read it you should read it through completely, without reference to the questions. Try and understand the information you have been given. It may be unfamiliar chemistry; that does not matter. Then read the questions and read the passage carefully again. Only then should you start to answer the questions.

An example of a comprehension question is given on page 80.

ASSESSMENT OBJECTIVES IN CHEMISTRY

Assessment Objectives are the intellectual and practical skills you should be able to show. Opportunities must be made by the Examiner when setting the examination paper for you to demonstrate your mastery of these skills when you answer the question paper.

Traditionally, the Assessment Objective of knowledge and understanding has been regarded as the most important skill to develop. Candidates have been directed to learn large bodies of knowledge to recall in the examination. Whilst not wanting in any way to devalue the learning of facts, it should be remembered that knowledge and understanding can only contribute about half of the marks on the written paper. The other half of the marks are acquired by mastery of the other Assessment Objectives. These are:

● To communicate scientific observations, ideas and arguments effectively.
● To select and use reference materials and translate data from one form to another.
● To interpret, evaluate and make informed judgements from relevant facts, observations and phenomena.
● To solve qualitative and quantitative problems.

The Periodic Table

KEY:

Atomic mass
Symbol
Name
Atomic number

s-block — Groups I, II
p-block — Groups III–VII, O
d-block — Transition elements
f-block

	I	II						Transition elements						III	IV	V	VI	VII	O
1	1 H Hydrogen 1																		4 He Helium 2
2	7 Li Lithium 3	9 Be Beryllium 4												11 B Boron 5	12 C Carbon 6	14 N Nitrogen 7	16 O Oxygen 8	19 F Fluorine 9	20 Ne Neon 10
3	23 Na Sodium 11	24 Mg Magnesium 12												27 Al Aluminium 13	28 Si Silicon 14	31 P Phosphorus 15	32 S Sulphur 16	35.5 Cl Chlorine 17	40 Ar Argon 18
4	39 K Potassium 19	40 Ca Calcium 20	45 Sc Scandium 21	48 Ti Titanium 22	51 V Vanadium 23	52 Cr Chromium 24	55 Mn Manganese 25	56 Fe Iron 26	59 Co Cobalt 27	59 Ni Nickel 28	64 Cu Copper 29	65 Zn Zinc 30	70 Ga Gallium 31	73 Ge Germanium 32	75 As Arsenic 33	79 Se Selenium 34	80 Br Bromine 35	84 Kr Krypton 36
5	85.5 Rb Rubidium 37	88 Sr Strontium 38	89 Y Yttrium 39	91 Zr Zirconium 40	93 Nb Niobium 41	96 Mo Molybdenum 42	99 Tc Technetium 43	101 Ru Ruthenium 44	103 Rh Rhodium 45	106 Pd Palladium 46	108 Ag Silver 47	112 Cd Cadmium 48	115 In Indium 49	119 Sn Tin 50	122 Sb Antimony 51	128 Te Tellurium 52	127 I Iodine 53	131 Xe Xenon 54
6	133 Cs Caesium 55	137 Ba Barium 56	139 La Lanthanum 57	178.5 Hf Hafnium 72	181 Ta Tantalum 73	184 W Tungsten 74	186 Re Rhenium 75	190 Os Osmium 76	192 Ir Iridium 77	195 Pt Platinum 78	197 Au Gold 79	201 Hg Mercury 80	204 Tl Thallium 81	207 Pb Lead 82	209 Bi Bismuth 83	210 Po Polonium 84	210 At Astatine 85	222 Rn Radon 86
7	223 Fr Francium 87	226 Ra Radium 88	227 Ac Actinium 89	104 Db Dubnium	105 Jl Joliotium	106 Rf Rutherfordium	107 Bh Bohrium	108 Hn Hahnium	109 Mt Meitnerium									

f-block

139 La Lanthanum 57	140 Ce Cerium 58	141 Pr Praseodymium 59	144 Nd Neodymium 60	147 Pm Promethium 61	150 Sm Samarium 62	152 Eu Europium 63	157 Gd Gadolinium 64	159 Tb Terbium 65	162.5 Dy Dysprosium 66	165 Ho Holmium 67	167 Er Erbium 68	169 Tm Thulium 69	173 Yb Ytterbium 70	175 Lu Lutetium 71
227 Ac Actinium 89	232 Th Thorium 90	231 Pa Protactinium 91	238 U Uranium 92	237 Np Neptunium 93	242 Pu Plutonium 94	243 Am Americium 95	247 Cm Curium 96	247 Bk Berkelium 97	251 Cf Californium 98	254 Es Einsteinium 99	253 Fm Fermium 100	256 Md Mendelevium 101	254 No Nobelium 102	257 Lr Lawrencium 103

Formulae, equations and amounts of substances

Each **element** is represented by a symbol. All of the symbols will be found in the Periodic Table (on page 6). Each **compound** is represented by a formula which gives the proportions of the different elements it contains. For example, 24 g of magnesium combines with 16 g of oxygen in every 40 g sample of magnesium oxide. Formulae of many compounds can be worked out using the following list of ions.

Common ions

Positive ions	Negative ions
lithium Li^+	fluoride F^-
sodium Na^+	chloride Cl^-
potassium K^+	bromide Br^-
silver Ag^+	iodide I^-
copper(II) Cu^{2+}	hydroxide OH^-
lead(II) Pb^{2+}	* nitrate NO_3^- (nitrate(V))
magnesium Mg^{2+}	* nitrite NO_2^- (nitrate(III))
calcium Ca^{2+}	carbonate CO_3^{2-}
strontium Sr^{2+}	hydrogencarbonate HCO_3^-
barium Ba^{2+}	* sulphate SO_4^{2-} (sulphate(VI))
zinc Zn^{2+}	hydrogensulphate HSO_4^-
aluminium Al^{3+}	* sulphite SO_3^{2-} (sulphate(IV))
iron(II) Fe^{2+}	oxide O^{2-}
iron(III) Fe^{3+}	sulphide S^{2-}
chromium(III) Cr^{3+}	* phosphate PO_4^{3-} (phosphate(V))
manganese(II) Mn^{2+}	manganate(VII) MnO_4^-
cobalt(II) Co^{2+}	ethanedioate $C_2O_4^{2-}$
nickel(II) Ni^{2+}	chlorate(V) ClO_3^-
ammonium NH_4^+	chlorate(I) OCl^-
** hydrogen H^+	chromate(VI) CrO_4^{2-}
	dichromate(VI) $Cr_2O_7^{2-}$

NB *Ions marked can be named in alternative ways. The systematic name is given in brackets, but the first name given is still preferred by IUPAC, national authorities and most Examination Boards. There are alternative names also for the parent acids, e.g. sulphuric acid (sulphuric(VI) acid), sulphurous acid (sulphuric(IV) acid), nitric acid (nitric(V) acid), nitrous acid (nitric(III) acid).**The hydrogen ion may be written as H^+ or in the hydrated form as H_3O^+ (oxonium ion) or $H^+(aq)$.

The following table shows how these ions can be used to obtain a formula.

Finding the formula

Compound	Ions present		Formula
copper(II) oxide	Cu^{2+}	O^{2-}	CuO
ammonium sulphate	NH_4^+	SO_4^{2-}	$(NH_4)_2SO_4$
potassium manganate(VII)	K^+	MnO_4^-	$KMnO_4$
calcium ethanedioate	Ca^{2+}	$C_2O_4^{2-}$	CaC_2O_4
chromium(III) oxide	Cr^{3+}	O^{2-}	Cr_2O_3
hydrochloric acid	H^+	Cl^-	HCl
nitric acid	H^+	NO_3^-	HNO_3
sulphuric acid	H^+	SO_4^{2-}	H_2SO_4

NB (i) All acids contain hydrogen ions (but see Chapter 16).
(ii) A small number after a bracket multiplies everything inside the bracket, e.g. $(NH_4)_2SO_4$ is composed of three ions – two NH_4^+ ions and one SO_4^{2-} ion. Overall there are two nitrogen atoms, eight hydrogen atoms, one sulphur atom and four oxygen atoms.

Working out a formula by experiment

There are a number of questions where you are expected to calculate a formula from data you are given.

E.g. *Calculate the formula of a hydrocarbon containing 2.4 g of carbon and 0.6 g of hydrogen.*

use the formula **number of moles = mass in g ÷ mass of 1 mole**

number of moles of carbon = 2.4/12 = 0.2

number of moles of hydrogen = 0.6/1 = 0.6

The simplest formula (often called the **empirical formula**) for the hydrocarbon is, therefore, CH_3. To find the actual **molecular formula** you would need to have the molecular mass or some information from mass spectra.

Calculate the empirical formula for the compound containing 40.1% carbon, 6.6% hydrogen and 53.3% oxygen.

This is a similar calculation to the last one if you consider a sample of 100 g of the compound. That sample contains 40.1 g of carbon, 6.6 g of hydrogen and 53.3 g of oxygen. If you use the relative atomic masses of carbon, hydrogen and oxygen as 12, 1 and 16, respectively, you should get an empirical formula of $C_1H_2O_1$.

Chemical equations

You will be expected to understand chemical equations using symbols and be able to write them correctly. The steps in writing a chemical equation are as follows:

❶ Write down the equation as a word equation. Include all of the reactants and products.

E.g. calcium carbonate + hydrochloric acid → calcium chloride + water + carbon dioxide

❷ Fill in the correct formulae for all of the reacting substances and products.

$CaCO_3 + HCl \rightarrow CaCl_2 + H_2O + CO_2$

❸ Now balance the equation. There must be the same total numbers of the different atoms before and after the reaction. When balancing an equation only the *proportions* of the reacting substances and products can be altered – not the formulae.

$CaCO_3 + 2HCl \rightarrow CaCl_2 + H_2O + CO_2$

❹ Finally, the states of the reacting substances and products can be included in small brackets after the formulae:

(s) for solid or (c) for crystalline,
(l) for liquid,
(g) for gas,
(aq) for a solution with water as solvent.

$CaCO_3(s) + 2HCl(aq) \rightarrow CaCl_2(aq) + H_2O(l) + CO_2(g)$

This equation also gives information about quantities reacting. The equation states that 1 mole of calcium carbonate reacts with 2 moles of hydrochloric acid to produce 1 mole of calcium chloride, 1 mole of water and 1 mole of carbon dioxide. We can turn these numbers of moles into masses using relative atomic masses, which are given in the Periodic Table.

1 mole of $CaCO_3 = 40 + 12 + (3 \times 16) = 100$ g
2 moles of $HCl = 2(1 + 35.5) = 73$ g
1 mole of $CaCl_2 = 40 + (35.5 \times 2) = 111$ g
1 mole of $H_2O = (2 \times 1) + 16 = 18$ g
1 mole of $CO_2 = 12 + (16 \times 2) = 44$ g

You will notice that the sum of the masses of the reactants equals the sum of the masses of the products.

You can now work out the mass of calcium chloride produced by 1g of calcium carbonate.

$$100 \text{ g CaCO}_3 \text{ produces } 111 \text{ g of CaCl}_2$$

$$1 \text{ g CaCO}_3 \text{ produces } \frac{111}{100} \text{ g of CaCl}_2 = 1.11 \text{ g}$$

You can work out the volume of 2M hydrochloric acid which would react with 1g of calcium carbonate.

2 moles of hydrochloric acid are present in 1000 cm³ of 2M hydrochloric acid (solution contains 2 mol dm⁻³)

1 g (= 0.01 moles) of CaCO₃ react with 10 cm³ of 2M hydrochloric acid.

You can also work out the volume of carbon dioxide produced at s.t.p. (273 K and 1 atm pressure) when 1g of CaCO₃ reacts with excess acid. To do this you use the information that 1 mole of gas occupies 22 400 cm³ at s.t.p.

100 g of CaCO₃ produces 1 mole of CO₂ which occupies 22 400 cm³ at s.t.p.

1 g of CaCO₃ produces 0.01 mole of CO₂ which occupies 224 cm³ at s.t.p.

Volumetric Calculations

These calculations involve reactions between volumes of solutions. Again this type of question uses a balanced symbolic equation.

E.g. $2\text{NaOH}(aq) + \text{H}_2\text{SO}_4(aq) \rightarrow \text{Na}_2\text{SO}_4(aq) + 2\text{H}_2\text{O}(l)$

Calculate the concentration of sulphuric acid (in mol dm⁻³) if 25 cm³ of sodium hydroxide (0.1 mol dm⁻³) is neutralised by 20 cm³ of sulphuric acid.

1000 cm³ of sodium hydroxide (0.1 mol dm⁻³) contains 0.1 moles of sodium hydroxide.

$$25 \text{ cm}^3 \text{ of } 0.1 \text{ mol dm}^{-3} \text{ sodium hydroxide contains } \frac{0.1 \times 25}{1000} \text{ moles}$$

$$= 0.0025 \text{ moles}$$

From the equation, 2 moles of sodium hydroxide react with 1 mole of sulphuric acid so 0.0025 moles NaOH react with 0.00125 moles H₂SO₄

0.00125 moles H₂SO₄ are present in 20 cm³ of sulphuric acid

$$\frac{0.00125 \times 1000}{20} \text{ moles of sulphuric acid would be present in 1000 cm}^3$$

Concentration of sulphuric acid = 0.0625 mol dm⁻³

Sometimes you will have to work out the volumes of gases at conditions other than s.t.p. You can use the formula

$$\frac{p_1V_1}{T_1} = \frac{p_2V_2}{T_2}$$

E.g. *Calculate the volume of carbon dioxide produced at 25 °C and 1.1 atm pressure.*
You know the volume at s.t.p. (273 K and 1 atm pressure) is 224 cm³. You wish to find the volume at 298 K and 1.1 atm pressure.

$$\frac{224 \times 1}{273} = \frac{1.1 \times V_2}{298}$$

$$V_2 = 222 \text{ cm}^3$$

1 *Formulae, equations and amounts of substances*

1   The mass of 1 mole of magnesium atoms is 24 g.
    What is the mass of one magnesium atom in grams?

    A 6.02×10^{23}
    B 6.02×10^{-23}
    C 3.99×10^{-23}
    D 3.99×10^{23} (1)

2   What is the minimum volume of 4 M hydrochloric acid required to dissolve 0.1 mole of
    magnesium according to the following equation?

    $$Mg + 2H^+ \rightarrow Mg^{2+} + H_2$$

    A $25\,cm^3$
    B $50\,cm^3$
    C $100\,cm^3$
    D $200\,cm^3$ (1)

                                                                                 SEB

3   The following half equations should be used to answer this question.

    $$Cr_2O_7^{2-}(aq) + 14H^+(aq) + 6e^- \rightarrow 2Cr^{3+}(aq) + 7H_2O(l)$$
    $$Fe^{2+}(aq) \rightarrow Fe^{3+}(aq) + e^-$$
    $$C_2O_4^{2-}(aq) \rightarrow 2CO_2(g) + 2e^-$$

    Which one of the following represents the correct stoichiometric ratio for the complete
    oxidation of iron(II) ethanedioate by dichromate(VI) ions?

    A $1Cr_2O_7^{2-} : 2FeC_2O_4$
    B $2Cr_2O_7^{2-} : 1FeC_2O_4$
    C $3Cr_2O_7^{2-} : 2FeC_2O_4$
    D $6Cr_2O_7^{2-} : 1FeC_2O_4$ (1)

4   A mixture of magnesium chloride and magnesium sulphate is known to contain 0.6 moles of
    chloride ions and 0.2 moles of sulphate ions. The number of moles of magnesium ions
    present is

    A   0.4
    B   0.5
    C   0.8
    D   1.0   (1)

                                                                                 SEB

5   Balance the following equations:

    (a) $Mn_3O_4 + Al \rightarrow Mn + Al_2O_3$
    (b) $NH_3 + O_2 \rightarrow NO + H_2O$
    (c) $Na_2H_2P_2O_7 + NaHCO_3 \rightarrow Na_2HPO_4 + CO_2 + H_2O$ (3)

6 Iron occurs in the Earth's crust as several different compounds, but only a few are used as ores from which iron is extracted. Two such compounds are the minerals haematite and magnetite, both of which are oxides of iron.

(a) 2.32 g of a pure sample of magnetite contained 1.68 g of iron. Calculate the formula of this oxide of iron.

....

.... (2)

(b) Iron is extracted from iron ores in the blast furnace. A mixture of the impure iron ore, carbon (in the form of coke), and calcium carbonate (in the form of limestone) is heated in the presence of a blast of air, and a series of reactions occurs. Three of these are:

$$2C + O_2 \rightarrow 2CO \qquad 1$$
$$CaCO_3 \rightarrow CaO + CO_2 \qquad 2$$
$$CO_2 + C \rightarrow 2CO \qquad 3$$

(i) If the iron oxide present is Fe_2O_3, write an equation, including state symbols, for the reaction between this compound and carbon monoxide to produce molten iron and carbon dioxide.

.... (2)

(ii) Suggest what further reaction is likely to happen to the carbon dioxide formed in this reaction.

.... (1)

The function of the calcium oxide formed in reaction 2 is to react with the impurities in the iron ore. The main impurity is silicon(IV) oxide, which combines with the calcium oxide in this reaction:

$$CaO + SiO_2 \rightarrow CaSiO_3$$

(iii) What is the name of the compound formed?

.... (1)

(iv) Write down the formula for the anion in this compound, including the ionic charge.

.... (1)

(c) Sand for glassmaking must be as free as possible from iron compounds as impurities. Sand is made of silicon(IV) oxide, SiO_2, and iron is usually present as a thin coating of iron(III) oxide, Fe_2O_3, on each grain of sand.

Use your knowledge of chemistry to suggest a method of removing the iron oxide coating from the sand grains that might be applied on an industrial scale. Justify your answer.

....

....

.... (3)

ULEAC Nuffield

7 An element X forms compounds XCl_2 and XBr_2. The dibromide is completely converted to the dichloride when it is heated in a stream of chlorine:

$$XBr_2 + Cl_2 \rightarrow XCl_2 + Br_2$$

When 1.500 g of XBr_2 is treated, 0.890 g of XCl_2 is formed. Calculate the relative atomic mass of X and, using the Periodic Table, identify the element X.

di.f => 0.61

....10.98 x 0.89 = 9.7..

.... (3)

8 Bronze is an alloy containing copper and tin. The percentage of tin in the alloy was determined by the method described.

15.0 g of finely powdered bronze was warmed with excess dilute sulphuric acid to convert all the tin to tin(II) sulphate. The mixture was then filtered to remove unreacted copper, and the colourless filtrate was made up to 250 cm³ with distilled water.

25.0 cm³ portions of the tin(II) sulphate solution were titrated against 0.0200 M potassium manganate(VII) solution. 28.0 cm³ of potassium manganate(VII) was needed to oxidize the tin(II) to tin(IV). The ionic equation for the reaction of tin(II) with manganate(VII) ions is:

$$5Sn^{2+} + 2MnO_4^- + 16H^+ \rightarrow 5Sn^{4+} + 2Mn^{2+} + 8H_2O$$

(a) Give the change in oxidation number of manganese and hence state whether it has been oxidized or reduced.

.... (1)

(b) Calculate the percentage by mass of tin in the alloy.

....

....

....

.... (4)

Oxford

Atoms are composed of **protons**, **neutrons** and **electrons**. The table compares the properties of these particles.

Name of particle	Mass	Charge	Place in the atom
proton, p	1 amu	1	in the nucleus
neutron, n	1 amu	0	in the nucleus
electron, e	negligible	−1	around the nucleus

amu = atomic mass unit

An atom is neutral and contains equal numbers of protons and electrons.

Atomic number (Z) is the number of protons in an atom.

Mass number (A) is the total number of protons plus neutrons in an atom.

Atoms of the same element but containing different numbers of neutrons are called **isotopes**.

E.g. A chlorine-35 atom contains 17 protons, 17 electrons and 18 neutrons, while a chlorine-37 atom contains 17 protons, 17 electrons and 20 neutrons.

Electrons are constantly moving around the nucleus. They are arranged in electron shells or energy levels. Each energy shell can only hold a certain maximum number of electrons.

Energy level or shell	Maximum number of electrons
1st	2
2nd	8
3rd	18
4th	32
5th, etc.	50

Electrons within an energy shell are placed into orbitals. Each orbital can hold a maximum of two electrons. Orbitals within a shell have slightly different energies and different shapes. Orbitals can be s, p, d and f. The diagram shows the order in which orbitals are filled.

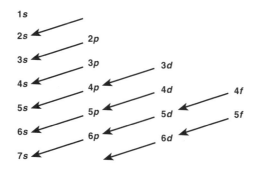

The electron arrangements of the first 20 elements are given on the next page. You should be able to work these out using the Periodic Table.

Electron arrangements of the first 20 elements

Hydrogen	$1s^1$
Helium	$1s^2$
Lithium	$1s^2 2s^1$
Beryllium	$1s^2 2s^2$
Boron	$1s^2 2s^2 2p_x^1$
Carbon	$1s^2 2s^2 2p_x^1 2p_y^1$
Nitrogen	$1s^2 2s^2 2p_x^1 2p_y^1 2p_z^1$
Oxygen	$1s^2 2s^2 2p_x^2 2p_y^1 2p_z^1$
Fluorine	$1s^2 2s^2 2p_x^2 2p_y^2 2p_z^1$
Neon	$1s^2 2s^2 2p_x^2 2p_y^2 2p_z^2$
Sodium	$1s^2 2s^2 2p_x^2 2p_y^2 2p_z^2 3s^1$
Magnesium	$1s^2 2s^2 2p_x^2 2p_y^2 2p_z^2 3s^2$
Aluminium	$1s^2 2s^2 2p_x^2 2p_y^2 2p_z^2 3s^2 3p_x^1$
Silicon	$1s^2 2s^2 2p_x^2 2p_y^2 2p_z^2 3s^2 3p_x^1 3p_y^1$
Phosphorus	$1s^2 2s^2 2p_x^2 2p_y^2 2p_z^2 3s^2 3p_x^1 3p_y^1 3p_z^1$
Sulphur	$1s^2 2s^2 2p_x^2 2p_y^2 2p_z^2 3s^2 3p_x^2 3p_y^1 3p_z^1$
Chlorine	$1s^2 2s^2 2p_x^2 2p_y^2 2p_z^2 3s^2 3p_x^2 3p_y^2 3p_z^1$
Argon	$1s^2 2s^2 2p_x^2 2p_y^2 2p_z^2 3s^2 3p_x^2 3p_y^2 3p_z^2$
Potassium	$1s^2 2s^2 2p_x^2 2p_y^2 2p_z^2 3s^2 3p_x^2 3p_y^2 3p_z^2 4s^1$
Calcium	$1s^2 2s^2 2p_x^2 2p_y^2 2p_z^2 3s^2 3p_x^2 3p_y^2 3p_z^2 4s^2$

Note: sometimes electronic structures are shown in a slightly condensed form, e.g. calcium $1s^2 2s^2 2p^6 3s^2 3p^6 4s^2$.

When an atom in its normal state (or ground state) is excited, electrons can be promoted to higher energy levels. When the atom is restored to its ground state, the extra energy is emitted in the form of a **spectrum**. This explains, for example, the characteristic orange flame colour of sodium.

The **mass spectrometer** is an instrument used for measuring atomic mass accurately, for finding the number of isotopes of an element present and for identifying atoms or groups of atoms present in a molecule.

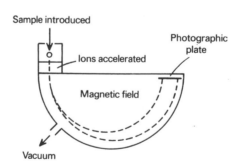

If you need to revise this subject more thoroughly, see the relevant topics in the *Letts A level Chemistry Study Guide*.

The vaporized sample being tested is introduced into the instrument and then is ionised by heating, electrical discharge or electron bombardment. The positive ions produced are accelerated by an electric field and passed through a slit to give a fine beam of ions all moving with the same velocity. The individual ions in the beam will differ slightly in mass and charge. The beam of ions is then subjected to a magnetic field which bends the beam into a circular path. Depending upon the mass and charge, the radius of the circular path for each ion is slightly different. The lighter the ion or the greater the charge on the ion the greater will be the deflection. The ions are detected by marks made on a photographic plate. You will find a number of examples in this book where a mass spectrometer has been used to solve problems.

1 Which statement **cannot** be true of two atoms with the same mass number?

      A   They are isotopes of the same element.
      B   They have different numbers of protons.
      C   They have different numbers of neutrons.
      D   They are atoms of two different elements.   (1)

      SEB

2 Which of the following particles will be formed when an atom of $^{211}_{83}\text{Bi}$ loses an alpha particle and the decay product then loses a beta particle?

      A $^{210}_{79}\text{Au}$

      B $^{209}_{80}\text{Hg}$

      C $^{209}_{81}\text{Tl}$

      D $^{207}_{82}\text{Pb}$ (1)

      SEB

3 A sample of oxygen consisting mainly of the isotope oxygen-16 was enriched with oxygen-18. The composition of the mixture was 75.0% oxygen-16 and 25.0% oxygen-18, by volume. The oxygen sample reacted with sodium as follows:

$$4\text{Na}(s) + \text{O}_2(g) \rightarrow 2\text{Na}_2\text{O}(s)$$

(a) Complete the table to show the composition of some of the species involved in the reaction. (A_r: ^{16}O, 16.0; ^{18}O, 18.0; ^{23}Na, 23.0)

species	protons	neutrons	electrons
$^{23}_{11}\text{Na}$			
$^{16}_{8}\text{O}$			
$^{18}_{8}\text{O}^{2-}$			

    (4)

(b) Write down the electronic configuration of:
(i) a sodium atom,
(ii) an oxide ion.

sodium atom ....

oxide ion .... (3)

(c) State **three** physical properties of sodium oxide.

....

....

.... (3)

(d) Calculate the relative atomic mass of oxygen in the sample above.

(2)

(e) Some carbon-12 was burned in another sample of the oxygen mixture. The carbon dioxide produced gave the following mass spectrum.

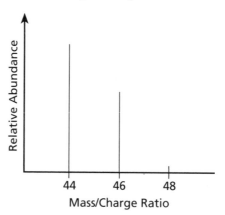

Identify the peaks shown.

....

....

.... (2)

UCLES

4 (a) Identify and give the main characteristics of the particles contained in atomic nuclei.

....

....

.... (2)

(b) Chlorine is essentially a mixture of two isotopes, ^{35}Cl and ^{37}Cl.
Explain what is meant by the term *isotopes*.

....

.... (2)

(c) The following is the mass spectrum of chlorine, Cl_2.

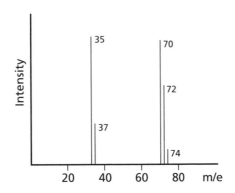

(i) Identify the species responsible for each peak in the spectrum.

....

....

.... (4)

(ii) Why are the peaks not all the same height?

....

.... (1)

(d) Complete the following equations for nuclear reactions:

(i) $^{9}_{4}Be$ + $^{4}_{2}He$ → $^{12}_{6}C$ + ...

(ii) $^{238}_{92}U$ → $^{234}_{90}Th$ + ...

(iii) $^{14}_{6}C$ → $^{14}_{7}N$ + ...

In (i), how many grams of $^{12}_{6}C$ would be produced from 1 g of $^{4}_{2}He$?

.... (4)

(e) State what is meant by the term *half-life* of a radioactive isotope.

....

.... (1)

ULEAC

5 A radioisotope used in a hospital has a half-life of 1.5 hours. It has a count rate of 8000 counts min^{-1} at 9.00 a.m.

(a) What would be the count rate at 1.30 p.m. on the same day?

.... (1)

(b) An aqueous solution of a compound containing the radioisotope was prepared. What effect would this have on its half-life?

.... (1)

(c) Give a use for radioisotopes in medicine.

.... (1)

SEB

6 Use the graph of first ionization energies of the elements from Na to Ar to answer the following questions.

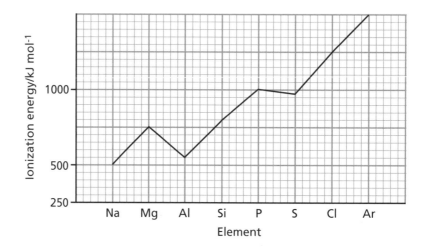

(a) Mark on the graph the approximate value of the ionization energy of potassium, K. (1)

(b) Explain the variation of first ionization energy in descending a group.

....

.... (2)

(c) Explain how the graph provides evidence for electron arrangements in *s* and *p* levels.

....

.... (2)

Oxford

7 The diagram shows the mass spectrum of a simple molecule A. Use the Periodic Table to answer the questions below.

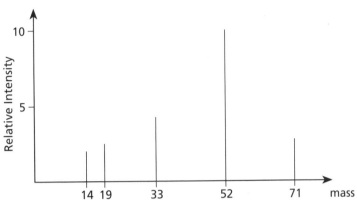

(a) What is the relative molecular mass of A?

.... (1)

(b) Calculate the difference in mass between the peak at 52 and the relative molecular mass. Hence suggest what atom or atoms could have been detached from A to produce the peak at 52.

.... (3)

(c) Calculate the difference in mass between the peak at 33 and the relative molecular mass, and similarly suggest what atom or atoms could have been detached from A to produce the peak at 33.

.... (2)

(d) Calculate the difference in mass between the peak at 14 and the relative molecular mass, and similarly suggest what atom or atoms could have been detached from A to produce the peak at 14.

.... (2)

(e) Suggest what the peak at mass 14 might represent, bearing in mind the absence of peaks of lower mass. Include the charge of the ion producing the peak.

....

.... (3)

(f) From your answers to the questions above, suggest what A might be.

....

.... (2)

AEB

3 Bonding and structure

Ionic bonding exists in compounds involving a metal and a non-metal, e.g. sodium chloride.

$$Na \longrightarrow Na^+ + e^- \qquad\qquad Cl + e^- \longrightarrow Cl^-$$
$$1s^2 2s^2 2p^6 3s^1 \qquad 1s^2 2s^2 2p^6 \qquad\qquad 1s^2 2s^2 2p^6 3s^2 3p^5 \qquad 1s^2 2s^2 2p^6 3s^2 3p^6$$

A **complete transfer** of one electron from a sodium atom to a chlorine atom produces sodium and chloride **ions**. Both sodium and chloride ions have stable noble gas electron arrangements. These ions are held together by strong electrostatic forces within the **lattice**. Compounds with ionic bonding usually have a high melting point (and high lattice energy). These compounds usually dissolve in water to produce a solution containing free ions.

 Covalent bonding involves a **sharing** of electrons. For example, in a molecule of methane, CH_4, the carbon atom is joined to four hydrogen atoms by single covalent bonds. The diagram shows a simple 'dot and cross' diagram for methane.

$$
\begin{array}{cc}
\text{H} & \text{H} \\
\text{O×} & | \\
\text{H ×O C O× H} & \text{H} - \text{C} - \text{H} \\
\text{×O} & | \\
\text{H} & \text{H}
\end{array}
$$

In each covalent bond two electrons, one from the carbon and one from the hydrogen, form an electron pair which holds the carbon and hydrogen atoms together. Covalent bonds are directional, giving definite shape to a molecule.

 Other examples of covalent bonding include:

$$\overset{××\ \ OO}{\underset{××\ \ OO}{\text{×Cl × Cl O}}} \qquad \overset{××\ \ \ O\,O}{\underset{×\times}{\text{×O ×O O}}}_{O} \qquad \overset{\ \ \ \times O}{\underset{\times O}{\text{× N ×O N}}}_{O}^{O}$$

$$Cl - Cl \qquad\qquad O = O \qquad\qquad N \equiv N$$

$$
\begin{array}{cc}
\text{H} & \text{H} \\
\text{O×} & \text{×O} \\
\text{×C ×O C} & \text{O} \\
\text{×O} & \text{×O} \\
\text{H} & \text{H}
\end{array}
\qquad\qquad
\begin{array}{c}
\text{H} \diagdown \qquad \diagup \text{H} \\
\qquad \text{C} = \text{C} \\
\text{H} \diagup \qquad \diagdown \text{H}
\end{array}
$$

In all the examples of covalent bonding given above, discrete molecules are produced. Substances containing molecules with covalent bonding will be gases or low boiling point liquids at room temperature. They will be insoluble in water but soluble in organic solvents.

 It is possible for covalent bonding to produce an extremely large network of atoms called a **giant structure**, e.g. diamond and silicon(IV) oxide. Substances with this type of structure will have extremely high melting points.

 Coordinate bonding (or dative covalency) is similar to covalent bonding, but both electrons are supplied by the same atom.

 E.g. the compound formed by the reaction of ammonia and boron trifluoride.

$$
\begin{array}{ccc}
\text{F} & & \text{H} \\
| & & | \\
\text{F} - \text{B} & \leftarrow \overset{O}{\underset{O}{}} \text{N} - & \text{H} \\
| & & | \\
\text{F} & & \text{H}
\end{array}
$$

The nitrogen atom supplies two electrons (the lone pair) which are then used to hold the boron and nitrogen atoms together. There are many examples of coordinate bonding in the formation of transition metal complexes.

Hydrogen bonding is more commonly between molecules (intermolecular) rather than within molecules (intramolecular). It occurs in compounds whose molecules contain a hydrogen atom covalently bonded to an electronegative atom, usually F, O or N. There are slight charges within the molecules caused by a slight shift of electrons in the covalent bonds. Weak electrostatic forces exist between the molecules.

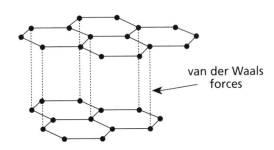

Hydrogen bonding has no effect on chemical properties but can alter physical properties such as boiling point and viscosity.

The weakest forces between molecules are called **van der Waals forces**.

E.g. graphite

van der Waals forces

Shapes of simple covalent molecules can be explained by the **electron-pair repulsion theory**. For example, a water molecule consists of an oxygen atom bonded to two hydrogen atoms by single covalent bonds. The shape of the molecule could, at first sight, be linear

i.e.   H−O−H

However, this would be ignoring the effect of the two lone pairs of non-bonding electrons on the oxygen atom. The four pairs of electrons (two covalent bonds and two non-bonding lone pairs) arrange themselves tetrahedrally around the oxygen atom. They do this to get as far apart as possible to reduce repulsion forces. Because the forces of repulsion between two non-bonding pairs of electrons is greater than other repulsion forces, the H−O−H bond angle is reduced from about 109° (the tetrahedral angle) to about 104°.

If you need to revise this subject more thoroughly, see the relevant topics in the *Letts* A level *Chemistry Study Guide*.

1   The bonding in a crystal of iodine is best described as

      A   covalent
      B   covalent and van der Waals
      C   ionic
      D   ionic and van der Waals
      E   van der Waals

(1)

NICCEA

2   Carbon dioxide is a gas at room temperature while silicon dioxide is a solid because

      A   van der Waals forces are much weaker than covalent bonds
      B   carbon dioxide contains double covalent bonds and silicon dioxide contains single covalent bonds
      C   carbon–oxygen bonds are less polar than silicon–oxygen bonds
      D   the relative formula mass of carbon dioxide is less than that of silicon oxide

(1)

SEB

3   The following table gives the electronegativities of some elements:

Element	Electronegativity	Element	Electronegativity
bromine	2.8	nitrogen	3.0
hydrogen	2.1	chlorine	3.0
carbon	2.5	oxygen	3.5

Which one of the following bonds would be expected to be most polar?

      A   C–H
      B   N–O
      C   N–Cl
      D   Br–Br
      E   H–Br

(1)

4   **Multiple completion**
Use the key below

A	B	C	D	E
(i), (ii) & (iii) only	(i) and (iii) only	(ii) and (iv) only	(iv) alone	(i) and (iv) only

Which of the following molecules contain(s) a bond angle which is smaller than the bond angle in CH_4? 109.5

    (i) SF_6 ⊃90°
    (ii) PF_5 ⊃90°
    (iii) H_2O – 104.5°
    (iv) NH_4^+ 109.5. ×

(1)

NEAB

5 (a) Complete the table of molecular shapes.

Number of pairs of bonding electrons	Number of lone pairs of electrons	Shape of molecule
3	0	Trig. Planer
3	1	Trig Pyramidal
4	0	Tetrahedralt
2	2	Bent.

(4)

(b) State briefly the meaning of the term *electronegativity*.

Measure of Strength of attraction for e̅

an atom has — + used for predicting Polarity. (1)

WJEC

6 The structures of solid caesium chloride, diamond, graphite, ice, iodine and sodium chloride are shown in the figure, but only that of ice is labelled.

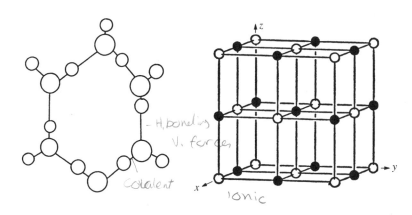

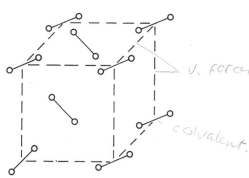

ice ....Na Cl............ ...Iodine............

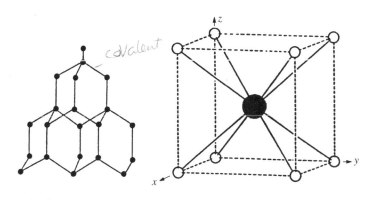

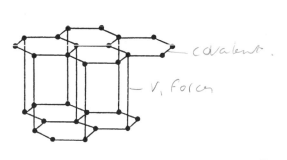

...Diamond......... ...S Caesium chloride - graphite......

23

(a) Label the remaining structures in the places provided. (4)

(b) Label each structure to show the **type** or **types** of bond present; if more than one type of bond is present show clearly **where one** example of **each** type is found in the structure. (6)

(c) List the six solids in approximate order of increasing melting point.

 ~~H₂O, Ceasium Cl, NaCl, BF₂, I₂ graphite Diamond~~

 Ice, I₂, NaCl, Caesium Cl

 diamond, & graphite (3)

WJEC

7 The graph shows the boiling points of the Group VI hydrides.

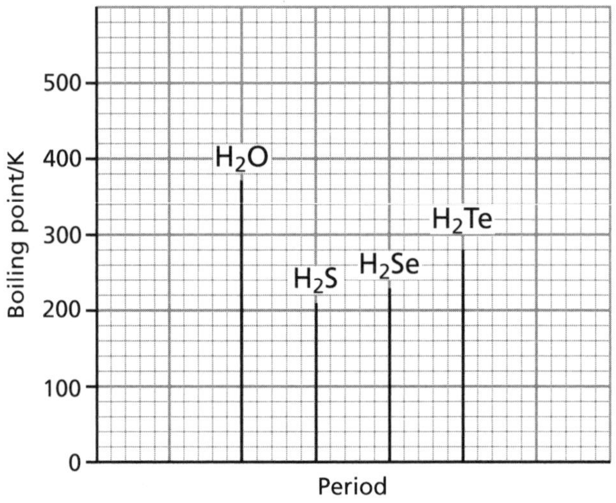

(a) Explain why the boiling points increase from H_2S to H_2Te.

 ....

 ....

 .... (2)

(b) Why does H_2O have an unusually high boiling point compared to other Group VI hydrides?

 ....

 .... (1)

SEB

8 Calcium fluoride is an ionic compound.

(a) (i) Indicate, using the axes below, the pattern of the first five ionisation energies for calcium.

Ionisation energy

Number of electrons removed

(2)

(ii) Explain the shape of your graph.

....

.... (2)

(b) The radius of a calcium atom is given in the *Book of Data* as 0.197 nm.
Which one of the following is the most likely value for the radius of a calcium ion?
Give a reason for your choice.

     0.100 nm   0.186 nm   0.197 nm   0.224 nm

Most likely value .... (1)

Reason .... (1)

(c) The ground state electronic configuration of an isolated fluorine atom is $1s^2 2s^2 2p^5$.

(i) Predict the ground state electronic configuration of a fluoride ion.

.... (1)

(ii) The radius of a fluoride ion (0.133 nm) is similar in value to the radius of a fluorine atom. What explanation can you give for this similarity?

....

.... (1)

(d) Explain how knowledge of the electronic structures of calcium and fluorine enables you to predict the formula for calcium fluoride. You may wish to use diagrams to assist your explanation.

....

....

.... (3)

ULEAC Nuffield

9 (a) The graph shows how the vapour pressure *p* of benzene varies with its temperature *T*.

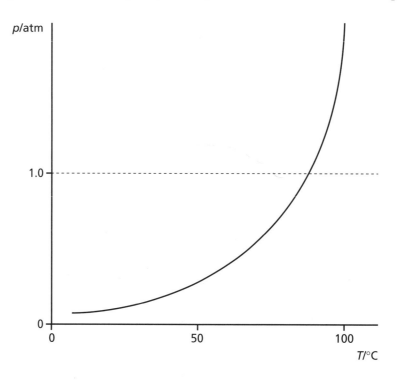

(i) Mark on the temperature axis with the letters b.p. the normal boiling point of benzene. (1)

(ii) By reference to the graph, explain how it is possible to maintain benzene as a liquid at temperatures higher than its normal boiling point.

....

.... (1)

(iii) Deduce from the graph the physical state of benzene at 75 °C and 0.3 atm.

.... (1)

(b) (i) Which property possessed by molecules is responsible for allowing the formation of a liquid?

.... (1)

(ii) Why, in terms of the energies of molecules, does increasing the temperature make it increasingly difficult to form a liquid?

....

.... (2)

(c) Using the axes below, sketch a distribution showing how the number N of molecules with given energy E in a gas varies with E at a given temperature.

(1)

NEAB

4 Energetics

Energy changes often accompany chemical reactions. A process which evolves energy is said to be **exothermic** and one which absorbs energy is **endothermic**. These energy changes result from changes in bonding during the reaction. Breaking bonds requires energy and forming bonds evolves energy. If more energy is produced from forming new bonds than is used to break bonds, the reaction is exothermic.

Tables of average bond energies (or enthalpies) enable you to calculate the expected overall enthalpy change of a reation by comparing the enthalpy required to break bonds and the enthalpy given out when bonds form.

Hess's law can be used to calculate enthalpy changes which cannot be determined by experiment.

E.g. *Calculate the enthalpy change of formation of carbon monoxide from the following information*:
Enthalpy changes of combustion of carbon(graphite) and carbon monoxide are $-393.5 \, kJ \, mol^{-1}$ and $-283.0 \, kJ \, mol^{-1}$.
The following equations can be written:
Required: Enthalpy change of formation of carbon monoxide (enthalpy change when 1 mole of carbon monoxide is formed from carbon(graphite) and oxygen).

$$C(graphite) \; + \; \tfrac{1}{2}O_2(g) \; \rightarrow \; CO(g) \; (\text{labelled } \Delta H_f^{\ominus})$$

The information you are given:
Enthalpy changes of combustion of graphite and carbon monoxide.

$$C(graphite) \; + \; O_2(g) \; \rightarrow \; CO_2(g) \qquad \Delta H^{\ominus} = -393.5 \, kJ$$

$$CO(g) \; + \; \tfrac{1}{2}O_2(g) \; \rightarrow \; CO_2(g) \qquad \Delta H^{\ominus} = -283.0 \, kJ$$

A diagram can be drawn to summarise the process:

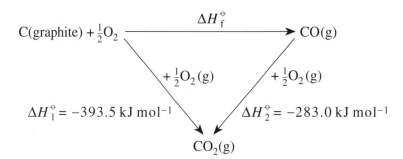

From this diagram we can complete the calculation.

$$\Delta H_f^{\ominus} = \Delta H_1^{\ominus} - \Delta H_2^{\ominus}$$

$$= -393.5 + 283.0$$

$$= -110.5 \, kJ \, mol^{-1}$$

The negative sign shows that the reaction is exothermic.

The **Born–Haber cycle** is used to determine lattice enthalpies indirectly. **Lattice energy** (or enthalpy) is the enthalpy change of formation of 1 mole of a solid ionic lattice from its free constituent ions in the gas phase.

The diagram summarises the Born–Haber cycle for sodium chloride:

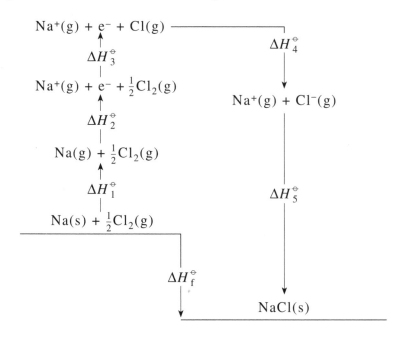

$\Delta H_1^\ominus$ corresponds to the **standard enthalpy change of atomisation** of sodium. It is the enthalpy change when 1 mole of solid sodium is converted into free gaseous sodium atoms.

$$\Delta H_1^\ominus = +108.4 \, \text{kJ mol}^{-1}$$

$\Delta H_2^\ominus$ corresponds to the **first ionisation energy of sodium**, i.e. the enthalpy change when 1 mole of electrons in their ground state is totally removed from 1 mole of sodium atoms in the gas phase.

$$\Delta H_2^\ominus = +500 \, \text{kJ mol}^{-1}$$

$\Delta H_3^\ominus$ corresponds to the **standard enthalpy change of atomisation** of chlorine. It is the enthalpy change when 1 mole of gaseous chlorine atoms is produced from chlorine molecules.

$$\Delta H_3^\ominus = +121 \, \text{kJ mol}^{-1}$$

$\Delta H_4^\ominus$ corresponds to the **first electron affinity** of chlorine. This is the enthalpy change when 1 mole of chlorine atoms accepts electrons to form 1 mole of chloride ions.

$$\Delta H_4^\ominus = -364 \, \text{kJ mol}^{-1}$$

$\Delta H_5^\ominus$ corresponds to the **lattice enthalpy** (sometimes called **lattice energy**) of sodium chloride.

$$\text{By Hess's law} \quad \Delta H_f^\ominus = \Delta H_1^\ominus + \Delta H_2^\ominus + \Delta H_3^\ominus + \Delta H_4^\ominus + \Delta H_5^\ominus = -411 \, \text{kJ mol}^{-1}$$
$$\Delta H_5^\ominus = -776.4 \, \text{kJ mol}^{-1}$$

Hess's law assumes that the compound is 100% ionic in bonding. If it is applied to a compound which is significantly covalent, such as cadmium iodide, the results will be inaccurate.

If you need to revise this subject more thoroughly, see the relevant topics in the *Letts* A level *Chemistry Study Guide.*

1 In which of the following changes is energy released?

> A $Cl_2(g) \rightarrow 2Cl(g)$
> B $NaCl(s) \rightarrow Na^+(g) + Cl^-(g)$
> C $Cl(g) + e^- \rightarrow Cl^-(g)$
> D $Na(g) \rightarrow Na^+(g) + e^-$ (1)

2 The enthalpy of combustion of methanol is -715 kJ mol^{-1}.
What mass of methanol, CH_3OH, has to be burned to produce 71.5 kJ?

> A 3.2 g
> B 32 g
> C 71.5 g
> D 715 g (1)

3 Consider the following bond enthalpies

Bond	Enthalpy/kJ mol^{-1}
Br–Br	194
H–Br	366
C–H	414
C–Br	280

What is the enthalpy change, in kJ mol^{-1}, for the following reaction?

$$
\begin{array}{ccccccc}
& H \quad H & & & & H \quad H & \\
& | \quad\; | & & & & | \quad\; | & \\
H-C-C-H & + & Br-Br & \rightarrow & H-C-C-Br & + & H-Br \\
& | \quad\; | & & & & | \quad\; | & \\
& H \quad H & & & & H \quad H &
\end{array}
$$

> A +38
> B −38
> C −1254
> D +1254 (1)

SEB

4 **Grid question**
There are a number of different enthalpy changes.

A combustion	B formation	C hydration
D neutralisation	E lattice-breaking	F sublimation

(a) Identify the enthalpy change which would be associated with the following reaction.

$$C(s) + 2H_2(g) \rightarrow CH_4(g)$$ (1)

(b) Identify the enthalpy change(s) which would be associated with the dissolving of an ionic salt in water. (1)

SEB

5 (a) This question is concerned with the reaction of hex-1-ene with hydrogen, giving hexane. The standard molar enthalpy changes of formation of hex-1-ene and hexane are -72.4 and -198.6 kJ mol^{-1}, respectively.

Define the term standard molar enthalpy change of formation.

....

.... (2)

(b) By drawing a suitable cycle, calculate the value of ΔH for the reaction hex-1-ene + hydrogen $\rightarrow$ hexane.

(3)

(c) Using the following average bond enthalpy terms: C–C 347, C–H 413, C=C 612 and H–H 436 kJ mol^{-1}, calculate a second value for ΔH for the reaction hex-1-ene + hydrogen $\rightarrow$ hexane.

....

....

.... (2)

(d) Which of your two answers do you think is more accurate, and why?

....

....

.... (3)

(e) The molar enthalpy change of formation of hexane cannot be measured directly, but its enthalpy change of combustion can. What other quantities would you need to measure to be able to calculate the enthalpy change of formation of hexane once you had measured its enthalpy change of combustion?

....

....

.... (2)

AEB

4 *Energetics*

6 (a) Draw and label carefully a Born–Haber cycle for the formation of calcium oxide.

(4)

(b) Use the data below to calculate the value of the lattice energies of calcium oxide and iron(II) oxide.

Standard enthalpy change of formation	(CaO) −635	(FeO) −278 kJ mol^{-1}
Standard enthalpy change of atomisation	(Ca) +178	(Fe) +416 kJ mol^{-1}
Standard molar 1st + 2nd ionisation energy	(Ca) +1735	(Fe) +2320 kJ mol^{-1}
Standard enthalpy change of atomisation	(O) +249 kJ mol^{-1} of O atoms	
Standard molar 1st + 2nd electron affinity	(O) +657 kJ mol^{-1} of O atoms	

....

....

....

.... (5)

(c) Suggest why the order of lattice energies is as given by your calculation.

....

.... (2)

(d) Calculate the enthalpy change for the reaction FeO + Ca → CaO + Fe. Do you think this might be a feasible method of making Fe on an industrial scale? Explain your answer.

....

.... (4)

(e) A student heats a mixture of calcium and iron(II) oxide in a test tube, under an inert atmosphere, but no reaction occurs. Suggest a reason for this.

....

.... (2)

(f) Calcium(III) oxide has never been formed, and its enthalpy of formation is thought to be positive.

   (i) Which terms in the Born–Haber cycle for calcium(III) oxide would be most different from those in the cycle for calcium(II) oxide, and how would they differ?

....

....

....

.... (4)

   (ii) What is the largest factor causing calcium(III) oxide to be endothermic?

.... (1)

(g) Iron(II) oxide can be oxidised readily to iron(III) oxide by warming it in oxygen. Explain why iron differs from calcium in this reaction.

....

....

.... (3)

AEB

7 A student investigated the enthalpy changes for reactions between a solution of silver nitrate, $AgNO_3$, and three different metal chloride solutions: NaCl(aq), KCl(aq) and $CaCl_2$(aq).

(a) The student first measured the standard enthalpy change for the reaction between silver nitrate and sodium chloride solutions in the following way.

$25.0 \, cm^3$ of $0.50 \, mol \, dm^{-3}$ silver nitrate solution was placed in a polystyrene foam cup with a close-fitting lid carrying an accurate thermometer.

$25.0 \, cm^3$ of $0.50 \, mol \, dm^{-3}$ sodium chloride solution was added, and thoroughly stirred in using the thermometer. This was repeated using fresh samples of each solution.

The temperature measurements were;

	Experiment 1	Experiment 2
Initial temperature: silver nitrate solution	17.6 °C	17.6 °C
sodium chloride solution	17.6 °C	17.7 °C
Final temperature of reaction mixture	21.1 °C	21.2 °C

   (i) Calculate the energy exchange between the reactants and the water in this experiment. Assume that all solutions require 4.17 J to change the temperature of $1 \, cm^3$ of solution by 1°C, and that there is no significant energy transfer to the apparatus or to the surroundings.

.... (1)

(ii) Use your answer to (i) to calculate a value for the standard molar enthalpy change for the reaction between silver nitrate and sodium chloride solutions.

....

.... (2)

(b) The student used previous chemical knowledge of reactions between ionic compounds in solution to predict the following standard enthalpy changes of reaction for each mole of metal chloride solution reacting:

A   The student predicted that the standard enthalpy change of reaction of silver nitrate solution with potassium chloride solution would have the same value as for the reaction with sodium chloride solution.

B   The student predicted that the standard enthalpy change of reaction of silver nitrate solution with calcium chloride solution would have twice the value as for the reaction with sodium chloride solution.

(i) Explain carefully how the student was able to make these predictions from previous chemical knowledge, giving chemical equations where appropriate.

....

....

....

.... (4)

(ii) The student decided to test these predictions experimentally. To test the first prediction the student used exactly the same procedure as for the sodium chloride solution, but using 0.50 mol dm^{-3} potassium chloride solution instead.

To test the calcium chloride solution the student considered whether, using the same concentration and volume of silver nitrate solution, any changes should be made to the concentration or volume of the chloride solution used.

What changes, if any, should the student make? Would the experimental results be affected? Justify your answer.

....

....

....

.... (4)

ULEAC

The factors which may affect the rate of a chemical reaction include:

**REVISION
SUMMARY**

❶ physical state of reactants;

❷ concentration (and for gases, pressure);

❸ temperature;

❹ catalysts;

❺ light.

The exact relationship between the rate of a reaction and the concentration of the reactants in a particular reaction can only be determined experimentally. Most reactions take place in a series of steps and the rate of the overall reaction depends upon the rate of the slowest step, called the **rate-determining step**.

It is important to distinguish between **order of reaction** and **molecularity**.

In the reaction

$$A + B \rightarrow C + D$$

you might find experimentally that the rate of reaction is proportional to the concentration of A and the square of the concentration of B. The rate equation can be written as:

$$\text{rate} = k[A][B]^2, \text{ where [A] and [B] represent the concentrations of A and B}$$

The **order of reaction** with respect to A is 1, i.e. $[A]^1$, and the order of reaction with respect to B is 2, i.e. $[B]^2$. The overall order is $1 + 2 = 3$. In the rate equation, k is called the **rate constant**.

Molecularity is the number of particles colliding in the rate-determining step. The order of reaction can be a fraction, but the molecularity must always be a whole number.

Arrhenius's equation is a relationship between rate constant and temperature.

$$k = Ae^{-E/RT} \text{ where } k \text{ is the rate constant}$$
$$A \text{ is the Arrhenius' constant}$$
$$E \text{ is the activation energy}$$
$$R \text{ is the gas constant}$$
$$T \text{ is the temperature (in kelvin)}$$

The activation energy can be found by plotting $\ln k$ against $1/T$. The gradient of the straight line graph is $-E/R$ and from this E can be found.

Catalysts are substances which alter the rate of a chemical reaction. An example is manganese(IV) oxide, which catalyses the decomposition of hydrogen peroxide.

$$2H_2O_2(aq) \rightarrow 2H_2O(l) + O_2(g)$$

A catalyst does not alter the position of an equilibrium or increase the yield of products. It provides an alternative route with a lower activation energy and so the reaction is speeded up. There are two types of catalyst – **heterogeneous** and **homogeneous**. A heterogeneous catalyst provides a surface at which the reaction can take place, e.g. finely divided iron provides a surface for the reaction of nitrogen and hydrogen gas molecules in the Haber process. In homogeneous catalysis the catalyst is in the same state as the reactants, e.g. acid-catalysed reactions such as the formation of an ester from ethanoic acid and ethanol.

If you need to revise this subject more thoroughly, see the relevant topics in the *Letts* A level *Chemistry Study Guide*.

1 When copper carbonate is reacted with excess acid, carbon dioxide is produced. The curves shown were obtained under different conditions.

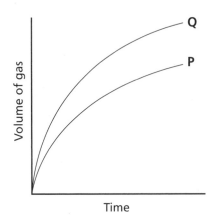

The change from P to Q could be brought about by:

    A increasing the concentration of the acid
    B decreasing the mass of copper carbonate
    C carrying out the reaction at a higher temperature
    D adding a catalyst   (1)

2 The rate equation for the reaction

$$C_2H_4(g) \ + \ H_2(g) \ \rightarrow \ C_2H_6(g)$$

is rate $= k[C_2H_4][H_2]$

If, at a fixed temperature, the reaction mixture is compressed to three times the original pressure, which one of the following is the factor by which the rate of reaction changes?

    A 3
    B 6
    C 9
    D 12
    E 27   (1)

NEAB

3 A reaction was carried out at four different temperatures. The table shows the times taken for the reaction to occur.

Temperature/°C	20	30	40	50
Time/s	60	30	14	5

The results show that:

    A a small rise in temperature results in a large increase in reaction rate
    B the activation energy increases with increasing temperature
    C the rate of reaction is directly proportional to the temperature
    D the reaction is exothermic   (1)

SEB

4 The Arrhenius equation accounts for the rate of a chemical reaction in terms of the

    A concentration of reactants
    B activation energy
    C physical states of the reactants
    D order of reaction   (1)

AEB

5

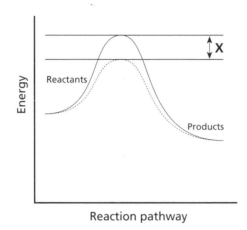

The diagram refers to the reaction

$$2SO_2 + O_2 \rightarrow 2SO_3$$

What does X represent?

    A Energy of activation for reaction without a catalyst
    B Enthalpy of reaction for reaction without a catalyst
    C Bond dissociation energy
    D None of these   (1)

6 Multiple completion

Use the key

A	B	C	D
1 & 2 only correct	2 & 3 only correct	1 only correct	3 only correct

For the reaction between A and B, the relationship between the concentrations of A and B and rate is given by

[A]/mol dm^{-3}	[B]/mol dm^{-3}	Relative rate
0.1	0.1	1
0.2	0.1	4
0.1	0.2	2

The reaction is

    1 first order with respect to A
    2 second order with respect to B
    3 third order overall   (1)

AEB

7 Ethyl ethanoate undergoes the following hydrolysis in alkaline conditions:

$$CH_3CO_2C_2H_5 + OH^- \rightarrow CH_3CO_2^- + C_2H_5OH$$

The following data were obtained from a series of experiments, conducted at constant temperature.

Initial rate of reaction (mol dm^{-3} s^{-1})	[CH$_3$CO$_2$C$_2$H$_5$] (mol dm^{-3})	[OH$^-$] (mol dm^{-3})
8.00×10^{-4}	6.4×10^{-2}	6.4×10^{-2}
8.00×10^{-4}	3.2×10^{-2}	12.8×10^{-2}
4.00×10^{-4}	1.6×10^{-2}	12.8×10^{-2}

(a) What is the order of the reaction with respect to:

(i) OH$^-$

.... (1)

(ii) $CH_3CO_2C_2H_5$?

.... (1)

(b) Write down the rate equation for the alkaline hydrolysis of ethyl ethanoate.

.... (1)

(c) Calculate the value of the rate (velocity) constant. Include the units.

....

....

....

.... (3)

Oxford

8 The thermal decomposition of ozone (trioxygen, O_3) is represented by the equation below and is second order with respect to ozone:

$$2O_3(g) \rightarrow 3O_2(g)$$

(a) Write down the rate expression for this reaction, and use it to define the terms 'rate constant' and 'order of reaction'.

....

....

.... (3)

(b) The value of the rate constant at 298 K is 3.38×10^{-5} dm^3 mol^{-1} s^{-1}.

(i) State the units in which the rate of the reaction was measured.

.... (1)

(ii) Calculate the rate of reaction at 298 K when the concentration of ozone, O_3, is 0.25 mol dm^{-3}.

(2)

(c) Explain why the rate of a reaction is increased by:

(i) increasing the temperature,

....

.... (2)

(ii) adding a catalyst.

....

.... (2)

(d) (i) **Outline** how the hydrolysis of bromoethane, C_2H_5Br, by aqueous sodium hydroxide could be shown to be first order with respect to bromoethane.

$$C_2H_5Br + NaOH \rightarrow C_2H_5OH + NaBr$$

....

....

....

.... (3)

(ii) Write down a mechanism for the reaction, given that it is also first order with respect to hydroxide ions.

....

.... (2)

ULEAC

Letts
Q&A

9 (a) The experimental rate equation for the reaction of the hydroxide ion, OH⁻, with $(CH_3)_3CBr$ is given by

$$\frac{-d[(CH_3)_3CBr]}{dt} = k[(CH_3)_3CBr]$$

where k is a constant

(i) State the order of the reaction.

.... (1)

(ii) What is the name of the constant, k?

.... (1)

(b) The reaction

$$C_6H_5N_2Cl(aq) + H_2O(aq) \rightarrow C_6H_5OH(aq) + N_2(g) + HCl(aq)$$

evolves nitrogen quantitatively at 40 °C.
Draw a clearly labelled diagram of the apparatus which you would use to monitor the evolution of nitrogen in this reaction at 40 °C.

(4)

(c) The following graph shows how the concentration, c, of $C_6H_5N_2Cl$ varies with time, t, in such an experiment.

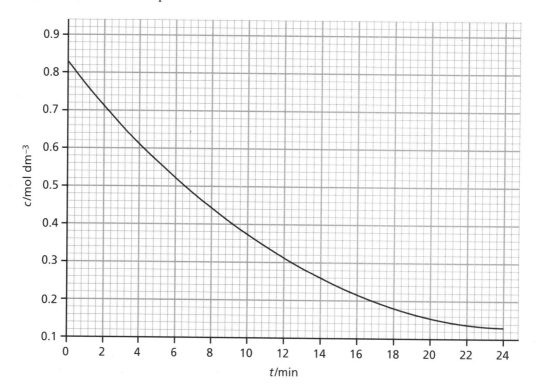

(i) Determine the rate of reaction when c is:

    (1) 0.300 mol dm^{-3}
    (2) 0.600 mol dm^{-3}

    (1) 0.300 mol dm^{-3}

    ....

    ....

    (2) 0.600 mol dm^{-3}

    ....

    .... (3)

(ii) Hence deduce the order of reaction with respect to $C_6H_5N_2Cl$.

    ....

    .... (1)

    Oxford

6 Equilibrium

In a **reversible reaction**, for example

$$A + B \rightleftharpoons C + D$$

it is possible to establish an **equilibrium** if the reaction is carried out in closed conditions so that products cannot escape.

In a dynamic equilibrium both the forward and the reverse reactions are taking place at the same rate. Consequently, the concentrations of A, B, C and D are not changing. The equilibrium is unstable and can be easily disturbed.

It is possible to predict changes which might take place to a system in equilibrium using **Le Chatelier's principle**.

$$\text{E.g.} \quad 2SO_2(g) + O_2(g) \rightleftharpoons 2SO_3(g) \quad \Delta H_{forward} \text{ is negative (exothermic)}$$

According to Le Chatelier's principle, the equilibrium would move to the right, producing more SO_3, by

❶ increasing the concentration of oxygen;

❷ removing SO_3 from the mixture;

❸ increasing the pressure, as there are fewer molecules on the right-hand side;

❹ lowering the temperature, as moving to the right produces more heat.

Lowering the temperature actually slows the rate of both the forward and reverse reactions.

For a reaction

$$aA + bB \rightleftharpoons cC + dD \text{ at constant temperature}$$

it is possible to write **equilibrium constants**:

$$K_c = \frac{[C]^c [D]^d}{[A]^a [B]^b}$$

where [A], [B], [C] and [D] are concentrations in mol dm^{-3},

or
$$K_p = \frac{p_C^c \, p_D^d}{p_A^a \, p_D^d}$$

where p_A, p_B, p_C and p_D are the partial pressures of the gases.

Brønsted and Lowry defined an **acid** as a proton donor and a **base** as a proton acceptor. **Lewis acids** are electron pair acceptors and **Lewis bases** are electron pair donors. Examples of Lewis acids and bases are boron trifluoride, BF_3, and trimethylamine, $(CH_3)_3N{:}$, respectively.

A weak acid is an acid which is only slightly ionised.

$$\text{E.g.} \quad HA \rightleftharpoons H^+ + A^-$$

Applying the principles of equilibrium, **Ostwald's dilution law** can be obtained

$$K = \frac{\alpha^2}{(1-\alpha)V}$$ where α is the degree of dissociation and V is the volume in dm^3 containing 1 mole.

A strong acid, e.g. hydrochloric acid, is completely dissociated.

$$HCl \rightarrow H^+ + Cl^-$$

The **ionic product of water** is derived from the ideas of equilibrium. It provides a relationship between the concentrations of hydrogen ions and hydroxide ions in an aqueous solution at constant temperature.

$$K_w = [H^+][OH^-] = 1 \times 10^{-14} \, mol^2 \, dm^{-6}$$

The **pH** of an aqueous solution is defined as $-\log_{10}[H^+]$. A neutral solution has $[H^+] = 10^{-7}$ mol dm^{-3} and has a pH of 7. A solution containing $[H^+] = 10^{-1}$ mol dm^{-3} has a pH of 1.

The ideas of equilibrium can be applied to the equilibrium between an almost insoluble compound and its ions in solution. This leads to the concept of **solubility product**, which explains the conditions under which precipitation from a solution will occur. For example, silver chloride, AgCl, is a sparingly soluble salt.

$$AgCl(s) \rightleftharpoons Ag^+(aq) + Cl^-(aq)$$

Applying the concept of equilibrium, the solubility product can be written as:

$$K_s = [Ag^+][Cl^-] = 2 \times 10^{-10} \, mol^2 \, dm^{-6} \text{ at } 25°C$$

Silver chloride will be precipitated from solution if the product of the silver ion concentration and chloride ion concentration exceeds 2×10^{-10} mol^2 dm^{-6}.

Solubility product can be used to calculate the solubility of a sparingly soluble solid in water. For example, the solubility product of iron(III) hydroxide, Fe(OH)$_3$, at 25°C is 8×10^{-40} mol^4 dm^{-12}. Calculate the solubility of iron(III) hydroxide in a saturated solution at 25°C. Relative molecular mass of iron(III) hydroxide = 107.

$$K_s = [Fe^{3+}][OH^-]^3 = 8 \times 10^{-40} \, mol^4 \, dm^{-12}$$
$$3[Fe^{3+}] = [OH^-]$$

Substitute
$$K_s = [Fe^{3+}](3[Fe^{3+}])^3$$
$$= 27[Fe^{3+}]^4 = 8 \times 10^{-40}$$
$$[Fe^{3+}]^4 = 0.3 \times 10^{-40}$$
$$[Fe^{3+}] = 0.74 \times 10^{-10} \, mol \, dm^{-3}$$

Mass of iron(III) hydroxide dissolving per dm^3 $= 0.74 \times 10^{-10} \times 107$
$$= 7.9 \times 10^{-9} \, g$$

A **buffer solution** is a solution whose pH remains unchanged even if slightly contaminated with acid or alkali. A buffer solution is made by mixing a weak acid and a salt of the acid or a weak alkali and a salt of the alkali.

Examples of buffer solutions include:
- ammonia solution and ammonium chloride (gives a pH greater than 7),
- ethanoic acid and sodium ethanoate (gives a pH less than 7).

If you need to revise this subject more thoroughly, see the relevant topics in the *Letts* A level Chemistry Study Guide.

1 For the reaction

$$W(g) + 2X(g) \rightleftharpoons 2Y(g) + 3Z(g)$$

which **one** of the following expresses the units of the equilibrium constant?

    A $mol\ dm^{-3}$
    B $mol^2\ dm^{-6}$
    C $mol^3\ dm^{-9}$
    D $mol^4\ dm^{-12}$ (1)

2 If equal volumes of equimolar solutions of the following pairs of substances were mixed, which pair would constitute a buffer solution?

    A ethanoic acid and methanoic acid
    B ammonia and methylamine
    C hydrochloric acid and sodium chloride
    D hydrochloric acid and sodium hydroxide
    E ammonia and ammonium chloride   (1)

NEAB

3 The extraction of an organic compound from an aqueous solution using $100\ cm^3$ of ethoxyethane could be most efficiently carried out by using

    A 1 portion of $100\ cm^3$
    B 2 portions of $50\ cm^3$
    C 4 portions of $25\ cm^3$
    D 5 portions of $20\ cm^3$ (1)

AEB

4 The concentration of $OH^-(aq)$ ions in a solution is $0.1\ mol\ dm^{-3}$.
What is the pH of this solution?

    A 8
    B 13
    C 14
    D 15   (1)

5 Multiple completion
Use the key below

A	B	C	D	E
(i), (ii) and (iii) only	(i) and (iii) only	(ii) and (iv) only	(iv) alone	(i) and (iv) only

A chemical company wishes to exploit the reaction:

$$X(g) \rightleftharpoons Y(g) + Z(g) \quad \Delta H \text{ is positive}$$

It is found that under a certain set of conditions the yield of products is poor. Which of the following changes would be likely to be investigated by the company with a view to improving the economics of the process?

    (i) recycle unchanged reactants
    (ii) increase the temperature
    (iii) attempt to find a suitable catalyst
    (iv) increase the pressure   (1)

NEAB

6 Grid question

An industrial gas mixture is produced by the catalytic steam reforming of methane.

$$CH_4(g) + H_2O(g) \rightleftharpoons CO(g) + 3H_2(g) \quad \Delta H = +206 \, kJ \, mol^{-1}$$

Identify the change(s) which would move the equilibrium to the right.

A increasing the temperature	**B** increasing the concentration of hydrogen	**C** increasing the pressure
D adding more catalyst	**E** decreasing the pressure	**F** decreasing the temperature

(1)

SEB

7 The graph below shows how the concentration of a reactant and product change during the course of a reaction.

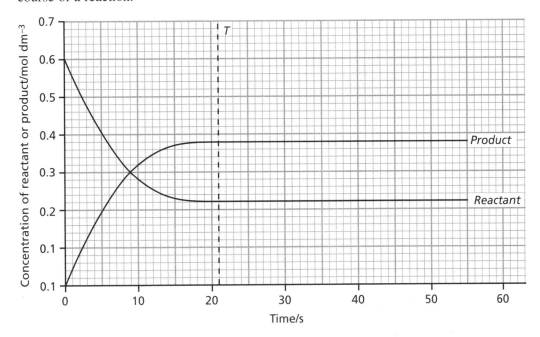

(a) Calculate the average rate of reaction over the first 15 s.

....

.... (1)

(b) After time *T* the concentrations of reactant and product remain constant.
What does this indicate about the reaction?

.... (1)

SEB

8 The table below gives information about the percentage yields of ammonia obtained in the Haber process under different conditions.

Pressure/atmospheres	Temperature/°C			
	200	300	400	500
10	50.7	14.7	3.9	1.2
100	81.7	52.5	25.2	10.6
200	89.1	66.7	38.8	18.3
300	89.9	71.1	47.1	24.4
400	94.6	79.7	55.4	31.9
600	95.4	84.2	65.2	42.3

(a) From the table, which combination of temperature and pressure gives the highest yield of ammonia?

.... (2)

(b) The equation for the main reaction in the Haber process is

$$N_2(g) + 3H_2(g) \rightleftharpoons 2NH_3(g) \quad (\Delta H_{forward} \text{ is negative})$$

Use this information to explain the effect of (i) pressure and (ii) temperature on the equilibrium yield of ammonia.

....

....

.... (2)

(c) In practice, the conditions used are 400 °C and 200 atmospheres. Explain why these conditions are used rather than those which give the highest yield.

....

....

.... (2)

SEB

9 Ammonia can be oxidised by air to form nitrogen oxide.

$$4NH_3(g) + 5O_2(g) \rightleftharpoons 4NO(g) + 6H_2O(g) \quad \Delta H = -909 \text{ kJ mol}^{-1}$$

This reaction forms the first stage in the manufacture of nitric acid from ammonia.

(a) State, and explain in terms of Le Chatelier's principle, the change in equilibrium yield of nitrogen oxide caused by:

(i) increasing the pressure at constant temperature;

....

.... (2)

(ii) increasing the temperature at constant pressure.

....

.... (2)

(b) The industrial process is operated at a temperature of about 900 °C. Bearing in mind your answer to (a)(ii), suggest a reason for this temperature.

....

.... (1)

(c) In the industrial process the mixture of gases is passed through gauzes of a platinum–rhodium alloy. This alloy acts as a *heterogeneous* catalyst.

(i) Suggest **one** reason for using the catalyst.

.... (1)

(ii) Explain the term *heterogeneous*.

.... (1)

(iii) Give **one** advantage of using a catalyst in a gauze form.

.... (1)

(d) If nitrogen oxide is mixed with air it can react to form a brown gas. This brown gas is an equilibrium mixture.

$$2NO_2(g) \rightleftharpoons N_2O_4(g)$$

At 77 °C and 700 kPa pressure, an equilibrium mixture contains 48% by volume of N_2O_4.

(i) Write the expression for the equilibrium constant, K_p, for this equilibrium. Include units in your expression.

.... (2)

(ii) Calculate the value of K_p at 77 °C.

....

....

.... (3)

AEB

10 (a) Comment on the following statements:

(i) A weak acid is one whose aqueous solution does not contain many $H^+(aq)$ ions.

....

.... (2)

(ii) A neutral solution is one which does not contain any $H^+(aq)$ or $OH^-(aq)$ ions.

....

.... (2)

(b) Calculate the pH values of the following solutions ($K_w = 1.00 \times 10^{-14}$ mol^2 dm^{-6} at 298 K):

(i) an aqueous HCl solution of concentration 0.01 mol dm^{-3};

....

.... (1)

(ii) aqueous KOH solution of concentration 0.02 mol dm^{-3};

....

.... (1)

(iii) aqueous Ba(OH)$_2$ solution of concentration 0.05 mol dm^{-3}.

....

.... (1)

(c) Describe how you would make a buffer solution, starting with pure ethanoic acid.

....

.... (2)

(d) Explain the meaning of the term *buffer solution*.

....

.... (2)

(e) Give an example of where a buffer solution would be of practical value.

.... (1)

AEB

11 (a) Define the term *Brønsted–Lowry acid*.

.... (1)

(b) Write an ionic equation for the reaction between ammonium chloride and sodium hydroxide and explain why this can be classified as an acid–base reaction.

Equation .... (1)

Explanation ....

.... (2)

(c) State what you understand by the term *strong* when applied to an acid.

.... (1)

(d) Define pH.

.... (1)

(e) Calculate the pH of 0.20 M aqueous hydrochloric acid.

....

.... (1)

(f) Given that at a particular temperature the ionic product of water is $1.00 \times 10^{-14} \, \text{mol}^2 \, \text{dm}^{-6}$, calculate the pH of 0.05 M aqueous sodium hydroxide.

....

....

....

.... (4)

(g) Hydrofluoric acid, HF, is a weak acid. Write an equation for the acid–base equilibrium established when this acid is added to water and derive an expression for the equilibrium constant, K_a, for this reaction.

Equation .... (1)

$K_a =$ .... (1)

NEAB

7 Redox

Oxidation is a process where electrons are lost and **reduction** where electrons are gained. A **reducing agent** is an electron donor and an **oxidising agent** is an electron acceptor. Oxidation and reduction occur together and may be called a **redox** reaction.

Common oxidising agents include:

- oxygen $O_2 + 4e^- \rightarrow 2O^{2-}$
- chlorine $Cl_2 + 2e^- \rightarrow 2Cl^-$
- iodine $I_2 + 2e^- \rightarrow 2I^-$
- manganate(VII) in acid solution $MnO_4^- + 8H^+ + 5e^- \rightarrow Mn^{2+} + 4H_2O$
- dichromate(VI) in acid solution $Cr_2O_7^{2-} + 14H^+ + 6e^- \rightarrow 2Cr^{3+} + 7H_2O$

Common reducing agents include:

- iron(II) salts $Fe^{2+} \rightarrow Fe^{3+} + e^-$
- thiosulphate $2S_2O_3^{2-} \rightarrow S_4O_6^{2-} + 2e^-$
- ethanedioates (oxalates) $C_2O_4^{2-} \rightarrow 2CO_2 + 2e^-$

Hydrogen peroxide can act as either an oxidising agent or a reducing agent, depending upon the conditions.

$$\text{OA} \quad H_2O_2 + 2H^+ + 2e^- \rightarrow 2H_2O$$
$$\text{RA} \quad H_2O_2 \rightarrow 2H^+ + O_2 + 2e^-$$

A balanced ionic equation can be written by combining two ionic half-equations

$$MnO_4^- + 8H^+ + 5e^- \rightarrow Mn^{2+} + 4H_2O$$
$$\times 5 \qquad\qquad 5Fe^{2+} \rightarrow 5Fe^{3+} + 5e^-$$
$$\text{Add} \quad MnO_4^- + 8H^+ + 5Fe^{2+} \rightarrow 5Fe^{3+} + Mn^{2+} + 4H_2O$$

The system of **oxidation states** (or **oxidation numbers**) gives a guide to the extent of oxidation or reduction in a species. The oxidation state can be defined as the number of electrons which must be added to a positive ion to get a neutral atom or removed from a negative ion to get a neutral atom. For covalent species it is assumed that the electrons in the covalent bond actually go to the atom which is most electronegative.

Rules about oxidation states are:

❶ The oxidation state of all elements uncombined is zero.

❷ The algebraic sum of the oxidation states of the elements in a compound is always zero.

❸ The algebraic sum of the oxidation states of the elements in an ion is equal to the charge on the ion.

❹ The normal oxidation states of hydrogen and oxygen in compounds are $+1$ and -2, respectively.

If, during a chemical reaction, the oxidation state of an element increases, the element is said to be oxidised, e.g. Mn(II) $\rightarrow$ Mn(VII) is oxidation.

When a metal rod is dipped into a solution of metal ions an equilibrium is set up. This can be summarised by

$$M(s) \rightleftharpoons M^{n+} + ne^-$$

Whether the metal acquires an overall positive charge or not depends upon the position of the equilibrium. A reactive metal has a equilibrium well over to the right-hand side.

A metal dipping into a metal salt solution is called a **half-cell**. Combining two half-cells produces an **electrochemical cell**. The diagram on the next page shows one such cell.

Letts
Q&A

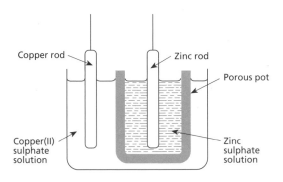

$$Zn(s) + Cu^{2+}(aq) \rightarrow Zn^{2+}(aq) + Cu(s)$$

The cell can be represented by

$$Zn(s) \mid Zn^{2+}(aq) \parallel Cu^{2+}(aq) \mid Cu(s)$$

The voltage of this cell is found by adding together the two **standard electrode potentials**.

$$Zn(s) \rightarrow Zn^{2+}(aq) + 2e^- \qquad E^\ominus = +0.76\,V$$

$$Cu^{2+}(aq) + 2e^- \rightarrow Cu(s) \qquad E^\ominus = +0.34\,V$$

$$\text{Voltage of cell} = +1.10\,V$$

The arrangement of elements in order of their standard electrode potentials leads to the **electrochemical series**. The elements with the greatest negative electrode potentials are at the top of the list.

decreasing reactivity →	**Metals**	**Non-metals**	increasing reactivity →
	potassium		
	sodium		
	magnesium		
	zinc		
	iron		
	copper		
	silver	iodine	
		bromine	
		chlorine	
		fluorine	

If a chemical reaction is to take place spontaneously, the **standard free energy change**, $\Delta G^\ominus$, must be negative. The equation which relates $\Delta G^\ominus$ and the standard redox potential $E^\ominus$ is

$$\Delta G^\ominus = -nFE^\ominus$$

where n is the number of electrons transferred and F is the Faraday constant. For a reaction to be feasible, $E^\ominus$ should be positive. A positive $E^\ominus$ value does not give any indication of the speed of the reaction.

If you need to revise this subject more thoroughly, see the relevant topics in the *Letts* A level *Chemistry Study Guide*.

7 Redox

1 In the reaction between an aqueous solution of bromine and a sodium sulphite solution, which of the following is oxidised?

    A OH^-
    B SO_3^{2-}
    C Na^+
    D Br_2 (1)

2 What is observed when chlorine gas is passed through a solution of potassium iodide containing starch?

    A the liquid turns dark blue
    B a yellow precipitate forms
    C a white precipitate forms
    D the liquid turns pale yellow   (1)

3 $NO_3^-(aq) + 4H^+(aq) + 3e^- \rightarrow NO(g) + 2H_2O(l)$

$$Zn(s) \rightarrow Zn^{2+}(aq) + 2e^-$$

These equations represent a reaction between nitric acid and zinc. How many moles of $NO_3^-(aq)$ are reduced by 1 mole of zinc?

    A $^2/_3$
    B 1
    C $^3/_2$
    D 2   (1)
        SEB

4 Iodate ions in the presence of acid oxidise sulphite ions to sulphate ions. Which one of the following equations represents this reaction?

    A $IO_3^- + 2SO_3^{2-} + 2H^+ \rightarrow I_2 + 2SO_4^{2-} + H_2O$

    B $4IO_3^- + SO_3^{2-} + 22H^+ \rightarrow 2I_2 + SO_4^{2-} + 11H_2O$

    C $2IO_3^- + SO_3^{2-} + 2H^+ \rightarrow I_2 + SO_4^{2-} + H_2O$

    D $2IO_3^- + 5SO_3^{2-} + 2H^+ \rightarrow I_2 + 5SO_4^{2-} + H_2O$

    E $2IO_3^- + SO_3^{2-} + 10H^+ \rightarrow I_2 + SO_4^{2-} + 5H_2O$ (1)
        NICCEA

5 Given the following data

    $Fe^{2+}(aq) + 2e^- \rightleftharpoons Fe(s)$ $E^{\ominus} = -0.4\,V$

    $Cu^{2+}(aq) + 2e^- \rightleftharpoons Cu(s)$ $E^{\ominus} = +0.4\,V$

Which one of the following is the standard e.m.f. in volts of the cell below?

    $Fe(s)\,|\,FeSO_4(aq)\,\|\,CuSO_4(aq)\,|\,Cu(s)$

    A -0.8
    B -0.4
    C 0.0
    D $+0.4$
    E $+0.8$ (1)
        NEAB

6 (a) Redox reactions can be used to produce electric currents.

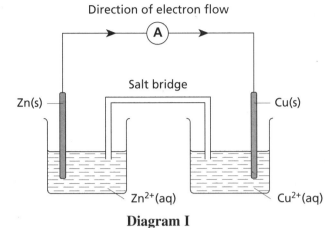

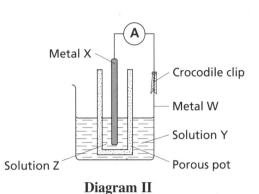

Diagram I   **Diagram II**

(i) The apparatus shown in diagram II may be used to carry out the **same** redox reaction as that described in diagram I. Identify the metal W and solution Z in diagram II.

.... (1)

(ii) What is the function of the salt bridge in I and the porous pot in II?

.... (1)

(iii) Write the ion–electron equation for the oxidation step.

.... (1)

(iv) The above reactions were carried out under standard conditions. What are the standard conditions for the above cells?

.... (1)

(v) If the ammeter were replaced by a voltmeter, what reading would be obtained?

.... (1)

Data from data sheet	
$Cu^{2+}(aq) + 2e^- \rightleftharpoons Cu(s)$	$+0.34\,V$
$Zn^{2+}(aq) + 2e^- \rightleftharpoons Zn(s)$	$-0.76\,V$

(vi) Explain why the concentration of copper(II) ions decreases when the cell is producing an electric current.

.... (1)

(b) Conversely, an electric current may be used to bring about redox reactions, using apparatus such as that shown below:

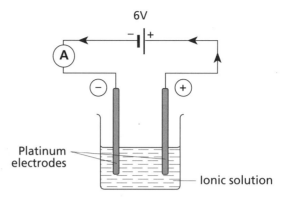

(i) Name a solution which would produce only hydrogen and oxygen at the platinum electrodes under standard conditions.

.... (1)

(ii) Draw a modified version of the above apparatus which would allow you to collect hydrogen and oxygen, indicating clearly which gas is collected at each electrode.

(2)

(c) A current of 0.4 A is passed through 1 M copper(II) sulphate solution. Calculate, to the nearest second, the time required to deposit 0.16 g of copper on the negative electrode.

....

.... (3)

SEB

7 A standard hydrogen electrode containing sulphuric acid and another half-cell consisting of a zinc rod dipping into aqueous zinc sulphate solution of concentration 1.00 mol dm^{-3} are linked by a salt bridge to form an electrochemical cell. The standard electrode potentials of the hydrogen and zinc half-cells are, respectively, 0.00 V and −0.76 V.

(a) (i) Write down the conventional representation of the cell.

.... (2)

(ii) What is the cell e.m.f. under standard conditions?

.... (1)

(iii) State and explain the effect on the cell e.m.f. of increasing the concentration of Zn^{2+} ions.

....

.... (2)

(b) (i) Write down the cell reaction which takes place when a current is drawn from the cell.

.... (2)

(ii) Which half-cell will be the positive pole of the cell?

.... (1)

(iii) Why are two half-cells required to make the complete cell?

....

.... (2)

(c) When a current is drawn from the cell the concentration of zinc ions in the zinc half-cell increases.
State and explain what change occurs in the pH in the hydrogen half-cell at the same time.

....

.... (2)

(d) Draw a fully labelled diagram to show how copper and zinc and appropriate solutions could be used to construct a simple storage cell. Indicate the direction of electron flow when a current is drawn.

(2)

ULEAC

8 (a) Explain why the plots of change of free energy, ΔG, with temperature, T, shown below for the reactions

$$2C(s) + O_2(g) \rightarrow 2CO(g) \qquad \Delta G = (-224 - 0.18T) \text{ kJ mol}^{-1}$$

$$C(s) + O_2(g) \rightarrow CO_2(g) \qquad \Delta G = (-394 - 0.00008T) \text{ kJ mol}^{-1}$$

have such gradients.

....

....

.... (3)

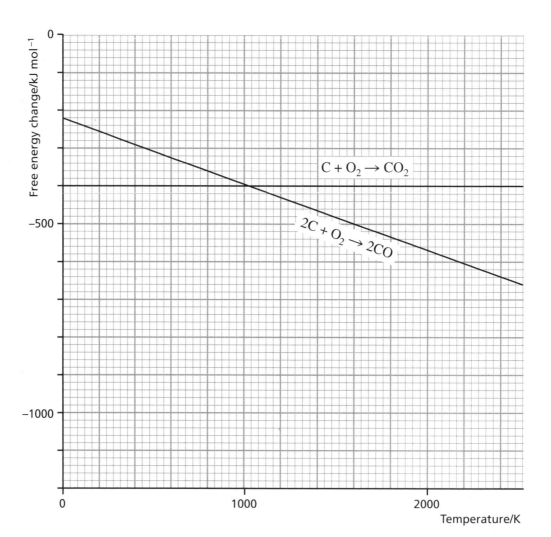

(b) For the reaction

$$\tfrac{4}{3}Al(s) + O_2(g) \rightarrow \tfrac{2}{3}Al_2O_3(s) \qquad \Delta G = (-1115 + 0.21T) \text{ kJ mol}^{-1}$$

predict graphically, using the grid in (a) above, the minimum temperature required for the reduction of alumina with graphite.

....

....

.... (3)

(c) Calculate the free energy change, ΔG, for the overall reaction that occurs at 1300 K for the extraction of aluminium in the Hall–Heroult cell.

$$\tfrac{1}{2}Al_2O_3 \; + \; \tfrac{3}{4}C \; \rightarrow \; Al \; + \; \tfrac{3}{4}CO_2$$

....

....

....

.... (4)

NEAB

8 *The Periodic Table*

The Periodic Table, first devised by Mendeleev in 1869, is an arrangement of all of the chemical elements in order of increasing atomic number. Elements with similar properties (i.e. the same chemical family) are placed in the same vertical column. The vertical columns are called groups and the horizontal rows are called periods.

The elements in Groups I and II are called *s*-block elements, Groups III–VII contain the *p*-block elements and the elements between Groups II and III are called *d*-block (or transition) elements.

There is a link between outer electron arrangement and the group in which the element is placed.

Element in Group	Outer electron arrangement
I	s^1
II	s^2
III	s^2p^1
IV	s^2p^2
V	s^2p^3
VI	s^2p^4
VII	s^2p^5
0	s^2p^6

There are trends in properties, such as melting points, atomic and ionic radii, electronegativity, ionisation energies, etc., across a period and down a group. These changes in properties are related to electron arrangements.

The graph shows the first ionisation energies of the elements plotted against atomic number. You will notice there are a series of peaks and troughs. Elements in the same group occupy similar positions on the graph, e.g. all of the Group I metals are at the lowest points of each trough.

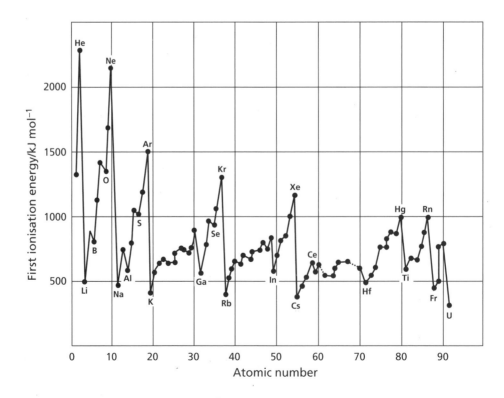

In a period of the Periodic Table there are trends in properties of compounds such as oxides and chlorides.

Oxides

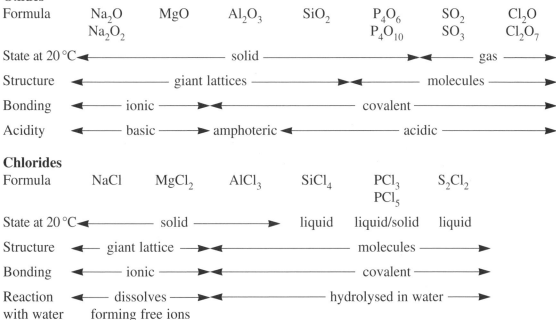

Formula	Na_2O	MgO	Al_2O_3	SiO_2	P_4O_6	SO_2	Cl_2O
	Na_2O_2				P_4O_{10}	SO_3	Cl_2O_7

State at 20 °C ◄————————— solid —————————► ◄—— gas ——►

Structure ◄———— giant lattices ————► ◄—— molecules ——►

Bonding ◄——— ionic ———► ◄————— covalent —————►

Acidity ◄——— basic ———► amphoteric ◄——— acidic ———►

Chlorides

Formula	NaCl	$MgCl_2$	$AlCl_3$	$SiCl_4$	PCl_3	S_2Cl_2
					PCl_5	

State at 20 °C ◄———— solid ————► liquid   liquid/solid   liquid

Structure ◄—— giant lattice ——► ◄————— molecules —————►

Bonding ◄——— ionic ———► ◄————— covalent —————►

Reaction ◄—— dissolves ——► ◄——— hydrolysed in water ———►
with water   forming free ions

The *d*-block elements, typified by the elements from scandium to zinc, are a block of metallic elements with similar properties. These elements are similar because the differences in electron arrangement are not in the outer electron shell but in inner shells. A *d*-block element has atoms which contain partially filled inner *d*-orbitals.

E.g. iron $1s^2 2s^2 2p^6 3s^2 3p^6 3d^6 4s^2$ (3*d* orbitals can hold a maximum of 10 electrons)

When iron loses two electrons to form an Fe^{2+} ion, the two electrons lost are the outer $4s^2$ electrons. The electron arrangement in an Fe^{2+} ion includes $3d^6$. An Fe^{3+} ion is formed when an iron atom loses three electrons (2 from 4*s* and 1 from 3*d*) and has a $3d^5$ arrangement.

The *d*-block elements have certain characteristic properties. These include:

❶ formation of compounds in a wide range of oxidation states;

❷ formation of coloured compounds;

❸ existence of paramagnetism; and

❹ formation of complexes.

Scandium and zinc do not possess these characteristics because they are not transition elements.

The table gives the possible oxidation states of the first row of *d*-block elements. The most common oxidation states are shown in bold type.

Element	Oxidation states					Examples
scandium	**+3**					Sc_2O_3
titanium	+2,	+3,	**+4**			TiO, Ti_2O_3, TiO_2
vanadium	+2,	+3,	**+4**,	+5		VO, V_2O_3, VO_2, V_2O_5
chromium	+2,	**+3**,	+6			CrO, Cr_2O_3, CrO_3
manganese	**+2**,	+3,	+4,	+6,	+7	MnO, Mn_2O_3, MnO_2, K_2MnO_4, $KMnO_4$
iron	+2,	**+3**				FeO, Fe_2O_3
cobalt	**+2**,	+3				CoO, Co_2O_3
nickel	**+2**,	+3,	+4			NiO, Ni_2O_3, NiO_2
copper	+1,	**+2**				Cu_2O, CuO
zinc	**+2**					ZnO

1 Which one of the following statements about the elements in Group I of the Periodic Table is **incorrect**?

    A   They all conduct electricity when molten or solid.
    B   Their chlorides are soluble in water.
    C   They are all reducing agents.
    D   They are produced at the cathode on electrolysis of aqueous solutions. (1)

2 Which one of the following lists of compounds consists ONLY of solids at room temperature which react with water to give acidic solutions?

A	P_4O_6	P_4O_{10}	Al_2Cl_6	PCl_5
B	SiO_2	P_4O_{10}	Al_2Cl_6	PCl_3
C	NaH	P_4O_6	$SiCl_4$	PCl_5
D	NaH	Cl_2O_7	P_4O_6	SO_3
E	SO_3	Cl_2O_7	Al_2Cl_6	PCl_3

    (1)

NICCEA

3–8 Classification questions

The outer electron arrangements of five transition elements are given below

		4s	3d
A	Chromium	↑	↑ ↑ ↑ ↑ ↑
B	Manganese	↑↓	↑ ↑ ↑ ↑ ↑
C	Iron	↑↓	↑↓ ↑ ↑ ↑ ↑
D	Copper	↑	↑↓ ↑↓ ↑↓ ↑↓ ↑↓
E	Zinc	↑↓	↑↓ ↑↓ ↑↓ ↑↓ ↑↓

Which of the transition elements:

3 shows an oxidation number of +1? .... (1)

4 shows an oxidation number of +7? .... (1)

5 has the smallest ionic radius? .... (1)

6 forms a colourless ion, M^{2+}? .... (1)

7 forms an ion M^{3+} with four unpaired $3d$ electrons? .... (1)

8 has the highest second ionisation energy? .... (1)

9 A sample of an orange-coloured solution goes yellow when alkali is added to it. Another sample of the solution goes green when sulphur dioxide is bubbled through it.

The orange-coloured solution contains

    A Cr^{3+}
    B CrO_4^{2+}
    C $Cr_2O_7^{2-}$
    D $[Fe(OH)_6]^{3-}$
    (1)

10 Multiple completion

Use the key

A	B	C	D
(i), (ii) and (iii) only	(i) and (iii) only	(ii) and (iv) only	(iv) alone

Examples in which the behaviour of lithium is not typical of group I are:

    (i) LiCl is soluble in water.

    (ii) LiCl has covalent character.

    (iii) Li reacts with water to form an alkaline solution.

    (iv) LiOH decomposes on heating.   (1)

NEAB

11 This question concerns the chlorides of the elements of Period 3 of the Periodic Table.

(a) In the table below write the formulae of the chlorides.   (4)

Na	Mg	Al	Si	P	S	Cl	Ar
						Cl_2	

(b) Select ONE of the chlorides above and outline a convenient laboratory method for its preparation.

....

....

....(3)

(c) Write equations for the changes which occur when sodium chloride and phosphorus trichloride are added to water.

    (i) Sodium chloride

    ....(1)

    (ii) Phosphorus trichloride

    ....(1)

(d) When an aqueous solution of magnesium chloride was evaporated to dryness, a product of composition Mg 35.5%, Cl 52.6%, O 11.9% by mass was obtained.

    (i) Calculate the empirical formula of the product.

    ....

    ....

    ....(2)

(ii) Why is anhydrous magnesium chloride not formed?

.... (1)

(e) (i) Give the electronic structure of the higher chloride of phosphorus in the gaseous state.

(1)

(ii) Sketch this molecule to show its shape, including the value(s) of the bond angles.

(1)

ULEAC

12 (a) (i) What is the characteristic feature present in the water molecule which enables it to act as a ligand?

.... (1)

(ii) Metal aqua ions contain two types of bond, M–O bonds and O–H bonds; breaking of either of these bonds involves a chemical reaction. Give one equation in each case to show a reaction involving the breaking of one of these bonds and name the type of reaction occurring. (You may use 'M' or any metal of your choice in these equations.)

M–O bond breaking equation ....

Reaction type ....

O–H bond breaking equation ....

Reaction type .... (4)

(b) The anti-cancer drug 'cisplatin' has the formula $[Pt(NH_3)_2Cl_2]$

(i) Give the oxidation state and coordination number of platinum in 'cisplatin'.

Oxidation state ....

Coordination number .... (2)

(ii) In aqueous solution in the body, one chloride ion in 'cisplatin' is substituted by a water molecule. Construct an equation for the reaction which occurs.

.... (2)

(c) Explain what is meant by the terms 'bidentate' and 'multidentate' as applied to ligands and give one example of each.

Explanation of bidentate ....

Example of bidentate ....

Explanation of multidentate ....

Example of multidentate .... (4)

NEAB

13 EDTA titrations can be used in quantitative analysis of many metal ions, such as nickel ions. The hydrated nickel ion is $[Ni(H_2O)_6]^{2+}$, and all six water molecule ligands can be replaced by **one** molecule of EDTA, giving a nickel–EDTA complex.

The relative molecular mass of a nickel salt can be determined by treating a known concentration of it with excess standard EDTA solution, in the presence of an alkaline buffer solution. The unreacted EDTA is then back-titrated with standard magnesium sulphate solution, in the presence of a little solochrome black.

(a) (i) What is the oxidation state of nickel in the $[Ni(H_2O)_6]^{2+}$ ion?

.... (1)

(ii) Draw a sketch of the hydrated nickel ion, naming its shape.

(2)

(b) (i) What effect does a buffer have?

.... (1)

(ii) Why is it necessary to add excess EDTA?

.... (1)

(iii) Suggest the purpose of adding solochrome black.

.... (1)

(c) Give **one** other use of EDTA.

.... (1)

Oxford

14 This question is about the elements in Group VII of the Periodic Table.

(a) Complete the following table (3)

Halogen	Physical state at room temperature	Colour
Fluorine	gas	yellow with greenish tinge
Chlorine		
Bromine		
Iodine		

(b) When potassium chloride is treated with concentrated sulphuric acid, a steamy gas, Y, is evolved.

(i) Write an equation for this reaction.

.... (1)

(ii) When Y is dissolved in water, in which it is readily soluble, a strongly acidic solution, Z, is formed. Describe the reaction that occurs between Y and water in terms of the Brønsted–Lowry theory.

....

.... (2)

(iii) Silver nitrate solution is added to Z. A white precipitate forms. If silver nitrate in concentrated aqueous ammonia is carefully added to Z, no precipitate forms. Explain these observations.

....

....

.... (3)

(c) Aqueous chlorate(I) ions, ClO^-, decompose on warming in a disproportionation reaction. Write the equation for this reaction and, by considering the oxidation state changes involved, explain the term 'disproportionation'

....

....

.... (3)

Organic chemistry is the study of the very wide range of compounds of carbon, excluding some simple compounds such as carbon dioxide, carbon monoxide and carbonates.

Organic compounds with the same molecular formula may have different structural or spatial arrangements. The existence of different compounds with the same molecular formula is called **isomerism** and the compounds are called **isomers**. Examples of different types of isomerism are give in the Questions section.

Reactions in organic chemistry can be classified into different groups.

● **Addition**. Two molecules combine together to form a single product.

E.g. (1)

$$H_2C=CH_2 + HBr \rightarrow CH_3CH_2Br$$

ethene + hydrogen bromide → bromoethane

(2)

$$CH_3CH_2CHO + HCN \rightarrow CH_3CH_2C(CN)(OH)H$$

propanal + hydrogen cyanide → propanal cyanohydrin

Reaction 1 is an example of **electrophilic addition**. The initial step is the formation of a weak π complex between the positive end of the hydrogen bromide molecule and the electrons in the π bond of ethene. The whole process can be summarised by:

$$H_2C=CH_2 \rightarrow H-CH_2-CH_2^+ + Br^- \rightarrow CH_3CH_2Br$$

Reaction 2 is an example of **nucleophilic addition**, with the initial attack by an electron-rich nucleophile, CN^-.

$$CH_3CH_2C(O^{\delta-})(H)^{\delta+} \xrightarrow{CN^-} CH_3CH_2C(O^-)(H)(CN) \xrightarrow{H^+} CH_3CH_2C(OH)(H)(CN)$$

Free radical substitution to alkenes is also possible, e.g. in polymerisation.

● **Substitution**. One group in a molecule is replaced by another.

E.g. $R–X + Y \rightarrow R–Y + X$

Substitution can be nucleophilic, electrophilic or free radical.

An example of nucleophilic substitution is:

$$CH_3CH_2I + OH^- \rightarrow CH_3CH_2OH + I^-$$
iodoethane + hydroxide ion → ethanol + iodide ion

Iodoethane is heated, under reflux, with an aqueous solution of potassium hydroxide (containing the OH^- nucleophile).

REVISION SUMMARY

Arenes, such as benzene, undergo electrophilic substitution reactions.

E.g. nitration

$$\text{benzene} + \text{HNO}_3 \rightarrow \text{nitrobenzene} + \text{H}_2\text{O}$$

benzene + mixture of conc. → nitrobenzene + water
nitric and sulphuric acids

The mixture of benzene and concentrated acids has to be heated but below 55°C. NO_2^+ is the electrophile.

An example of a free radical substitution is

$$CH_4 + Cl_2 \rightarrow CH_3Cl + HCl$$
methane + chlorine → chloromethane + hydrogen chloride

Chlorine free radicals (Cl•) are formed in small numbers in UV light (sunlight).

$$Cl\text{–}Cl \rightarrow 2Cl•$$
$$Cl• + CH_4 \rightarrow H\text{–}Cl + CH_3• \quad \text{(methyl free radical)}$$
$$CH_3• + Cl\text{–}Cl \rightarrow CH_3Cl + Cl•$$

The reaction continues as a **chain reaction**.

- **Elimination**. A molecule is split up into pieces (the opposite of addition).

E.g. $CH_3CH_2I + KOH \rightarrow H_2C{=}CH_2 + KI + H_2O$

Reflux iodoethane with potassium hydroxide (dissolved in ethanol). This is an elimination reaction.

- **Condensation**. Addition occurs, followed immediately by elimination. A condensation reaction occurs when two molecules join and a small molecule is eliminated (see Unit 10 – Condensation polymers).

E.g.

$$PhNHNH_2 + CH_3CH_2C\overset{O}{\underset{H}{\diagdown}} \rightarrow CH_3CH_2C\overset{NNHPh}{\underset{H}{\diagdown}} + H_2O$$

phenylhydrazine + propanal → propanal phenylhydrazone + water

If you need to revise this subject more thoroughly, see the relevant topics in the *Letts* A level Chemistry Study Guide.

- **Hydrolysis**. A molecule is split up with water. Usually, the splitting up occurs faster if dilute acid or alkali is used.

E.g. ethanenitrile $\quad CH_3CN + 2H_2O \rightarrow CH_3COONH_4 \quad$ (ammonium ethanoate)
$$H_2O + CH_3CN + NaOH \rightarrow CH_3COO^-Na^+ + NH_3 \quad \text{(sodium ethanoate)}$$
$$2H_2O + CH_3CN + HCl \rightarrow CH_3COOH + NH_4Cl \quad \text{(ethanoic acid)}$$

- **Oxidation–Reduction**

E.g. oxidation of ethanol $\quad CH_3CH_2OH + [O] \rightarrow CH_3CHO + H_2O \quad$ (ethanal)
$$CH_3CHO + [O] \rightarrow CH_3COOH \quad \text{(ethanoic acid)}$$

Heat ethanol and acidified potassium dichromate(VI) together. Colour goes from orange to green. The reverse reaction is reduction. Common reducing agents for this reduction are:

ethanoic acid to ethanol, lithium tetrahydridoaluminate(III) ($LiAlH_4$) in dry ether;

ethanal to ethanol, sodium amalgam and water.

1 The overall reaction

$$(CH_3)_2CHBr + NH_3 \rightarrow (CH_3)_2CH\overset{+}{N}H_3 + Br^-$$

is best described as

        A heterolytic fission
        B homolytic fission
        C nucleophilic substitution
        D electrophilic addition
        E cracking   (1)

NEAB

2 Which of the following compounds would undergo nucleophilic addition?

        A ethene C_2H_4
        B bromoethane C_2H_5Br
        C ethanal CH_3CHO
        D ethane C_2H_6 (1)

3 The main product of the addition of hydrogen bromide, HBr, to propene in an inert solvent in the dark is

        A 1-bromopropane
        B 1-bromopropene
        C 2-bromopropane
        D 2-bromopropene
        E 1,2-dibromopropane   (1)

4 Which one of the following compounds gives a proton NMR spectrum showing only one peak?

        A propan-1-ol
        B propan-2-ol
        C propanal
        D propanone   (1)

NEAB

Grid questions

5 Many organic compounds contain oxygen

A $CH_3-CH_2-\underset{\underset{H}{\mid}}{C}=O$	B $CH_3-CH_2-O-CH_3$	C $CH_3-O-\overset{O}{\overset{\|}{C}}-CH_3$
D $CH_3-\overset{O}{\overset{\|}{C}}-CH_3$	E $CH_3CH_2-C\overset{\diagup\,O}{\diagdown\,OH}$	F $CH_3-\underset{\underset{OH}{\mid}}{CH}-CH_3$

(a) Identify the ketone.   (1)

(b) Identify the compound which could be oxidised to form the compound in box E.   (1)

QUESTIONS

(c) Identify the compound which could be hydrolysed when warmed with sodium
hydroxide solution. F C . (1)

SEB

6

$$H-\underset{\underset{H}{|}}{\overset{\overset{H}{|}}{C}}-\underset{\underset{H}{|}}{\overset{\overset{H}{|}}{C}}-H \xrightarrow{process\ 1} \underset{H}{\overset{H}{>}}C=C\underset{H}{\overset{H}{<}} \xrightarrow{process\ 2} H-\underset{\underset{H}{|}}{\overset{\overset{H}{|}}{C}}-\underset{\underset{H}{|}}{\overset{\overset{H}{|}}{C}}-OH$$

A condensation	**B** cracking	**C** dehydration
D hydration	**E** hydrolysis	**F** oxidation

(a) Identify process 1. B (1)
(b) Identify process 2. D (1)

SEB

7 and 8 Multiple completion
Use the key

A	B	C	D	E
(i), (ii) & (iii) only	(i) and (iii) only	(ii) and (iv) only	(iv) alone	(i) and (iv) only

7 Which of the following give(s) a colour change on warming?

(i) CH_3CH_2OH and acidified potassium dichromate(VI)

(ii) $(CH_3)_3COH$ and acidified potassium dichromate(VI)

(iii) CH_3CH_2OH and Fehling's solution

(iv) CH_3CHO and Fehling's solution (1)

NEAB

8 Which of the following compounds exhibit(s) geometrical isomerism?

(i) $CH_3CH=CHCH_3$

(ii) $(CH_3)_2C=CH_2$ CH_3 · C

(iii) $(CH_3)_2C=C(CH_3)_2$

(iv) $CH_3CH=C(CH_3)CH_2CH_3$ (1)

NEAB

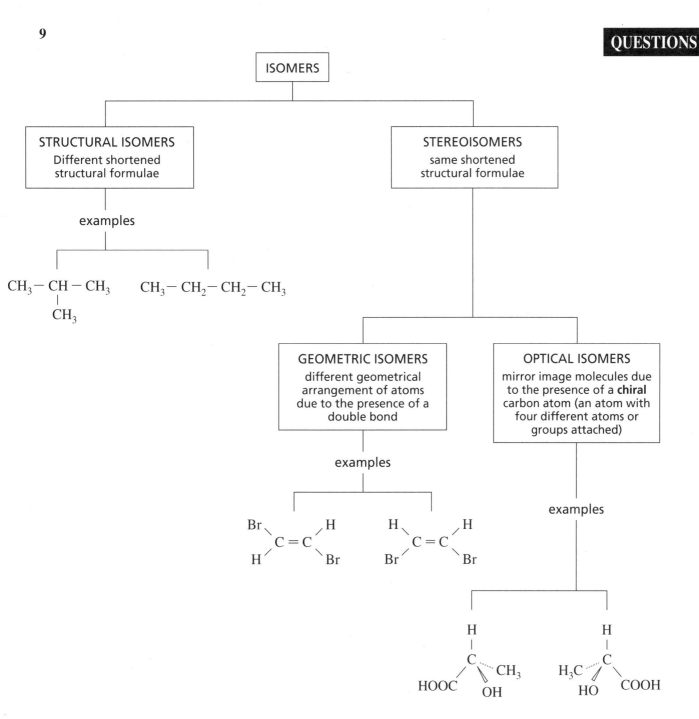

(a) Draw a structural isomer of 1,2-dibromoethane.

(1)

(b) Draw the geometric isomers of but-2-ene.

(1)

(c) Complete the diagram below to show the lightest alkane molecule containing a **chiral** carbon atom.

$$\overset{|}{\underset{\diagdown}{\text{C}}}$$

(1)

SEB

10 This question concerns two structural isomers

$$\text{CH}_3 \qquad\qquad\qquad \text{CH}_2\text{Br}$$

A   **B**

(a) Name each isomer

A ....

B .... (2)

(b) Describe, giving an outline of the mechanism and reaction conditions, how **A** and **B** could be prepared separately from methylbenzene.

A

B

(8)

(c) Describe a chemical test which would distinguish **A** and **B**.

....

....

....

.... (4)

(d) Outline a method which could be used to convert **B** into benzenecarboxylic acid (benzoic acid).

....

....

....

.... (6)

11 Each of the following concerns a pair of isomers. In each case write one possible structural formula for each of the species.

(a) **A** and **B** are alcohols with the molecular formula $C_4H_{10}O$. **A** turns warm acidifed potassium dichromate(VI) solution green, but **B** does not.

A   **B**

(2)

(b) **C** and **D** have the molecular formula C_4H_8O and neither contains a carbon–carbon double bond. **C** reacts with ammonical silver nitrate but **D** does not.

C   **D**

(2)

(c) **E** and **F** have the molecular formula $C_3H_6O_2$. **E** liberates carbon dioxide from sodium hydrogencarbonate but **F** does not.

E   **F**

(2)

(d) **G** and **H** are amines with the molecular formula C_7H_9N and both contain a benzene ring. **G** forms a stable diazonium salt but **H** does not.

G   **H**

(2)

Letts
Q&A

(e) **I** and **J** have the molecular formula C_4H_6. In a molecule of **I** all four carbon atoms lie in a straight line but in a molecule of **J** they do not.

I   **J**

(2)

NEAB

12 This question concerns compounds derived from ethane, C_2H_6.

(a) Parts of the infrared spectrum of ethanol, C_2H_5OH, and ethane are shown below:

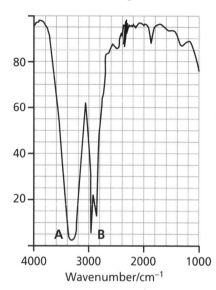   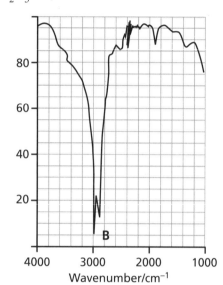

What chemical bonds do you think are responsible for the absorption of radiation at 3200–3400 cm^{-1} (Peak **A**) and 2950 cm^{-1} (Peak **B**).

Peak **A**: .... Peak **B**: .... (2)

(b) Suggest a route for the conversion of ethanol to ethylamine, stating the reactant(s) needed for each stage.

(3)

(c) The infrared spectrum of ethylamine is very similar to that of ethanol, including a peak of similar shape to that of peak **A** at almost the same wavenumber.

Suggest what group in ethylamine is responsible for this peak, and justify your answer.

....

.... (2)

ULEAC

13 (a) Phenylethanone ($C_6H_5COCH_3$) can be prepared from benzene.

(i) Give the reagents necessary to bring about this reaction.

.... (1)

(ii) Give the conditions necessary to bring about this reaction.

.... (1)

(iii) Write an equation for this reaction.

.... (1)

(iv) The reaction is an electrophilic substitution.
What is the electrophile?

.... (1)

(b) (i) Write an equation for the reaction between phenylethanone and hydrogen cyanide.

.... (1)

(ii) Give a mechanism for this reaction.

(2)

(iii) With the aid of diagrams explain why you would expect this product to be optically inactive.

(2)

(c) (i) Draw the structure of pentan-2-one.

(1)

(ii) Pentan-2-one forms two 2,4-dinitrophenylhydrazones which are geometric isomers. Draw the structures of these two isomeric phenylhydrazones.

(2)

ULEAC

10 *Some natural and synthetic materials*

Polymerisation is the forming of long chain **polymer** molecules from short **monomer** molecules. Polymers can be natural (starch, rubber, proteins) or synthetic (poly(ethene), nylon).

There are two types of polymerisation – **addition polymerisation** and **condensation polymerisation**. Poly(ethene) is an example of addition polymerisation. Ethene molecules (monomers) are joined by a series of addition reactions.

$$n \quad \overset{H}{\underset{H}{}} C = C \overset{H}{\underset{H}{}} \quad \rightarrow \quad \left[\begin{array}{cc} H & H \\ | & | \\ C & - C \\ | & | \\ H & H \end{array} \right]_n$$

The polymerisation can be carried out at high temperatures and pressures or at low temperatures using Ziegler catalysts ($(C_2H_5)_3Al$ and $TiCl_4$).

The table gives information about three other addition polymers.

Monomer	Structure of polymer	Name of polymer
chloroethene (vinyl chloride)	$\left[\begin{array}{cc} Cl & H \\ \| & \| \\ C & -C \\ \| & \| \\ H & H \end{array}\right]_n$	poly(chloroethene) (poly(vinyl chloride) or PVC)
phenylethene (styrene)	$\left[\begin{array}{cc} \bigcirc & H \\ \| & \| \\ C & -C \\ \| & \| \\ H & H \end{array}\right]_n$	poly(phenylethene) (polystyrene)
tetrafluoroethene	$\left[\begin{array}{cc} F & F \\ \| & \| \\ C & -C \\ \| & \| \\ F & F \end{array}\right]_n$	poly(tetrafluoroethene) (PTFE or Teflon)

Condensation polymerisation involves a series of condensation reactions, with a small molecule, such as water, hydrogen chloride or methanol, lost at each stage. The molecules forming a condensation polymer must contain two reactive groups. Terylene is a polyester and nylon a polyamide.

$$n\,HOCH_2CH_2OH \ + \ nCH_3O-\overset{O}{\overset{\|}{C}}-\bigcirc-\overset{O}{\overset{\|}{C}}-O-CH_3$$

ethane-1,2-diol   dimethyl benzene-1,4-dicarboxylate

$$\rightarrow \ \left[OCH_2CH_2O-\overset{O}{\overset{\|}{C}}-\bigcirc-\overset{O}{\overset{\|}{C}} \right]_n \ + \ 2n\,CH_3OH$$

Terylene

$$n\,NH_2(CH_2)_6NH_2 \ + \ nHOOC(CH_2)_4COOH$$

hexane-1,6-diamine   hexane-1,6-dioic acid
(hexamethylene diamine)   (adipic acid)

$$\rightarrow \ \left[NH(CH_2)_6NHCO(CH_2)_4CO \right]_n \ + \ 2n\,H_2O$$

Nylon 6,6

Proteins are natural condensation polymers made up from a series of amino acids. The amino acid molecules are joined by **peptide links**. These links can be broken to produce a mixture of amino acids that can be separated and identified by **chromatography**.

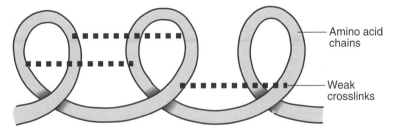

A single protein can contain as many as 500 amino acid units combined together. The sequence of amino acid units is complicated, as is the structure. A protein can be shown as coils, with loops of the coils held in position by weak crosslinks.

Amino acid chains

Weak crosslinks

With proteins and fats, **carbohydrates** make up the three major classes of nutrients. They are found in plants where they are produced from carbon dioxide and water by photosynthesis using solar energy.

Carbohydrates contain only carbon, hydrogen and oxygen and fit a formula of $C_x(H_2O)_y$. The simpler carbohydrates are sometimes called sugars. These are crystalline solids that dissolve in water to give a sweet-tasting solution. The simplest of these, including glucose and fructose, are **monosaccharides**, and are made up of six-carbon units. Glucose has a formula $C_6H_{12}O_6$ and behaves as an aldehyde. Like an aldehyde, it is a reducing agent and forms a red-brown precipitate when heated with Benedict's solution. It does not give a pink colour with Schiff's reagent and is not an aldehyde but is easily converted into one during reactions.

The modern structure of glucose is a hexagonal ring structure. Fructose has a pentagonal ring structure. Glucose solution usually contains an equilibrium mixture of α and β forms.

Disaccharides, e.g. sucrose and maltose, are formed when two monosaccharides join and eliminate water.

Polysaccharides are carbohydrates made from long chains of monosaccharide units. They are usually non-crystalline, generally insoluble in water and tasteless. They fit a formula $(C_6H_{10}O_5)_n$. Starch is a mixture of two polysaccharides, amylose and amylopectin, in a ratio of about 1 to 4. Both are made of glucose units, but joined in different ways. Amylose is responsible for the blue colour formed when starch reacts with iodine.

Hydrolysis of polysaccharides by boiling with dilute acid splits them up into their component monosaccharides by hydrolysis. The monosaccharides present can be identified by chromatography.

If you need to revise this subject more thoroughly, see the relevant topics in the *Letts A level Chemistry Study Guide*.

1 Which of the following descriptions is **not** true of proteins?
They are

      A natural condensation polymers
      B built up from identical monomer units
      C are able to be hydrolysed
      D built up from amino acids
                                                      (1)

2 A condensation polymer is classified as different from an addition polymer because the
condensation polymer is

      A a straight-chain hydrocarbon
      B re-softened on heating
      C formed by eliminating small molecules
      D formed by combination of monomers
                                                      (1)

SEB

3 This question is about the amino acid shown below

$$CH_3 - \overset{\displaystyle COOH}{\underset{\displaystyle NH_2}{\overset{|}{\underset{|}{C}}} } - H$$

Which one of the following will be the main species present in an aqueous solution of this
compound at pH14?

A $CH_3 - \overset{\displaystyle COOH}{\underset{\displaystyle NH_2}{\overset{|}{\underset{|}{C}}} } - H$

B $CH_3 - \overset{\displaystyle COOH}{\underset{\displaystyle NH_3^+}{\overset{|}{\underset{|}{C}}} } - H$

C $CH_3 - \overset{\displaystyle COO^-}{\underset{\displaystyle NH_3^+}{\overset{|}{\underset{|}{C}}} } - H$

D $CH_3 - \overset{\displaystyle COO^-}{\underset{\displaystyle NH_2}{\overset{|}{\underset{|}{C}}} } - H$

                                                          (1)

4 Sucrose, $C_{12}H_{22}O_{11}$, is a disaccharide which can be readily hydrolysed into glucose and fructose. The structure of sucrose is:

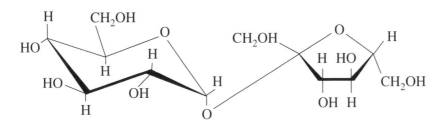

(a) What do you understand by the terms

(i) disaccharide ....

.... (1)

(ii) hydrolysed? ....

.... (1)

(b) Complete the structures for molecules of glucose and fructose by using the structure of sucrose.

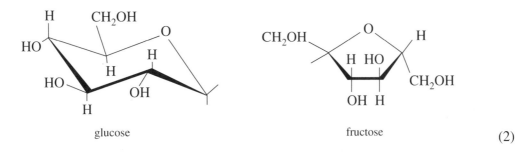

glucose   fructose   (2)

(c) In an experiment the hydrolysis of sucrose was investigated. 5 cm³ of concentrated hydrochloric acid was added to a solution of 100 g of (+)-sucrose in 40 cm³ of distilled water.
The mixture was placed in a polarimeter. The rotation of the plane of polarized light was measured at intervals over a 60-minute period.

α_1 represents the angle of rotation at time t.
α_2 is the final value of the angle of rotation.

time, t /minutes	α_1 /degrees	α_2 /degrees
0	+65	+80
10	+19	+34
20	+1	+16
30	−9	+6
40	−13	+2
50	−15	0
60	−15	(α_0)

Letts
Q&A

The mixture of glucose and fructose formed in this experiment is called 'invert sugar'.

(i) What can you say about the concentration of glucose and fructose in the solution of 'invert sugar' formed in this experiment? Give a reason for your answer.

....

.... (2)

(ii) Why is a solution of glucose optically active?

.... (1)

(iii) What is the meaning of the (+) and (−) signs in the table of results?

....

.... (2)

ULEAC

5 (a) *Ziegler catalysts* can be used to make *stereoregular polymers*.

(i) Give an example of a Ziegler catalyst.

.... (1)

(ii) What is meant by the term stereoregular polymer?

.... (1)

(b) When propene is polymerized using a Ziegler catalyst, both an atactic and an isotactic form of poly(propene) can be obtained from the mixture.

(i) Draw the structural formula of a section of the **isotactic** polymer chain. It should be made from **at least 4** monomer units.

(2)

(ii) Show, by reference to your answer in (i), how the polymer chain would differ in atactic poly(propene).

....

.... (1)

(iii) How does the structural type (i.e. isotactic or atactic) affect the properties of poly(propene).

....

.... (2)

(iv) Give an example of a use of isotactic poly(propene) and state how it is related to its structure.

....

.... (1)

Oxford

Comprehension question

ROCKET PROPELLANTS

Gunpowder

The Chinese used gunpowder as an explosive. They discovered that sometimes the gunpowder did not burn fast enough to explode. The powder, placed in a cardboard tube, burnt slowly and the tube started to fly along the ground. Thus rockets were invented and gunpowder was the first propellant.   5

Gunpowder consists of a mixture of finely divided potassium nitrate, sulphur and carbon. The proportions of the constituents and the main products of combustion correspond roughly with the following equation:

$$2KNO_3 + S + 3C \rightarrow K_2S + N_2 + 3CO_2$$

Carbon monoxide is also evolved and the residue contains potassium carbonate and   10
sulphate. Hence it is considered that the explosion of gunpowder cannot adequately be represented by a single chemical equation.

Current propellants

All current propellants are based upon a fuel and an oxidiser, which are pumped into a combustion chamber. Combustion is instantaneous, there is no need to ignite the   15
mixture!

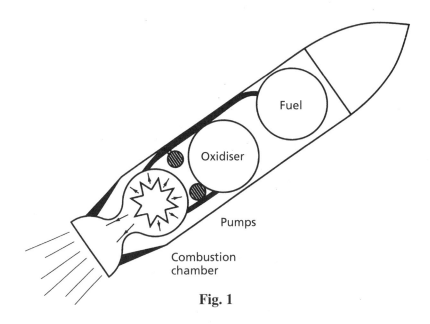

Fig. 1

There are three general properties that any good propellant must have. They are as follows:

❶ The reaction between the components must be highly exothermic, i.e. ΔH should
be large and negative.   20

❷ The reaction between the components must be fast.

❸ The products of the reaction should be gases of low molar mass. The result is a
stream of hot fast-moving gases which is ejected through the exhaust nozzle.

Despite the vast number of combustion reactions, the number of chemical propellants
in common use is relatively small. Examples of liquid propellants are shown in the table   25
on the next page:

Oxidiser	Fuel
nitrogen tetroxide	hydrazine
nitrogen tetroxide	unsymmetrical dimethylhydrazine
red fuming nitric acid	unsymmetrical dimethylhydrazine
chlorine trifluoride	hydrazine
liquid oxygen	kerosene (paraffin)
liquid fluorine	liquid hydrogen
liquid fluorine	hydrazine

The Saturn V rocket assembly, which placed men on the moon, used liquid oxygen   35
as an oxidiser together with kerosene and hydrogen as the fuels.

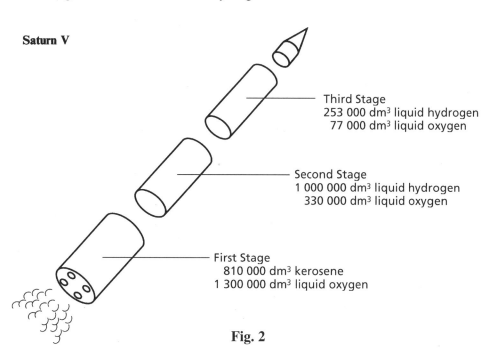

Saturn V

Third Stage
253 000 dm³ liquid hydrogen
77 000 dm³ liquid oxygen

Second Stage
1 000 000 dm³ liquid hydrogen
330 000 dm³ liquid oxygen

First Stage
810 000 dm³ kerosene
1 300 000 dm³ liquid oxygen

Fig. 2

Solid propellants are often more complex than their liquid counterparts. Ingredients
are generally classified according to their function, e.g. fuel, binder, curing agent,
burn-rate catalyst.

Ammonium perchlorate is the most widely used solid oxidiser. All of the perchlorate   40
oxidisers produce hydrogen chloride in their reactions with fuels. Their exhaust gases are
toxic and corrosive, to the extent that care is required in firing rockets, particularly the
very large ones, to safeguard operating personnel and communities in the path of the
exhaust cloud.

FUTURE PROPELLANTS

Free radicals   45

To date, the energy source used by rocket engines is the breaking and reforming of
chemical bonds. This source affords energy densities of approximately 13 kJ/g in the
liquid hydrogen/liquid oxygen combinations, up to 24 kJ/g with a lithium/fluorine
combination.

Theoretically, supplementary energy can be added to molecules or molecular   50
fragments that, upon recombination or relaxation to their normal energy state, release
significant amounts of energy. For example, 218 kJ/g is theoretically released when two
hydrogen atoms (free radicals) recombine to form hydrogen. Free radicals have been
stored in inert matrices at very low temperatures.

Excited atoms   55

Even higher energy densities are theoretically available from lightweight atoms, such as helium. Excited helium has a theoretical energy level of 479 kJ/g above the ground state. Assuming release of this energy and subsequent expansion through a rocket nozzle, this would produce a specific impulse greater than previously achieved. Techniques for generating activated helium and other noble gases are well known, but concentrating and   60
storing these species is difficult as they revert to their ground state by collision processes.

Metallic hydrogen

The creation and use of metallic hydrogen (hydrogen derived from normal hydrogen subjected to very high pressure) should release 218 kJ/g on the transition from the metallic to the normal solid form. The concept dates back to 1935, but interest has been   65
renewed because some scientists believe that metallic hydrogen exists in some large planets.

Answer all of the following questions

A1 Suggest what is meant by the following terms which are found in the passage:
   (a) free radical (line 53),
   (b) ground state (line 57),
   (c) burn-rate catalyst (line 39).

A2 Gunpowder burns with a lilac (purple) flame.
   (a) Which substance in gunpowder causes this colour (lines 6–7)?
   (b) Describe how you would prove the presence of carbonate in the residue from burnt gunpowder (line 10).

A3 (a) Draw the structures of symmetrical and unsymmetrical dimethylhydrazine.
   (b) Suggest a systematic name for unsymmetrical dimethylhydrazine (line 29).

A4 Explain why the reaction of hydrogen with fluorine is regarded as an oxidation reaction despite the fact that no oxygen is involved (line 33).

A5 (a) Deduce the formula of ammonium perchlorate (line 40).
   (b) Explain why the exhaust gases from propellants using perchlorate oxidisers are corrosive (line 42).

A6 Explain, using energy levels and electron configurations, what is meant by the term *excited helium* (line 57).

A7 What electrical properties would you expect metallic hydrogen to possess and, using accepted theories, how would you explain them (line 63)?

A8 The first stage of the Saturn V rocket involves the reaction of oxygen with kerosene. Assuming the formula of kerosene is $C_{12}H_{26}$, write a balanced equation for the complete combustion of kerosene (line 36).

A9 Calculate the volume of the gaseous exhaust in the third stage of the Saturn V rocket. Assume that the exhaust is composed completely of steam and is produced at a pressure of 1 atmosphere and a temperature of 3007 °C (Fig. 2).
   ($pV = nRT$; $R = 0.082$ atm dm^{3}K^{-1}mol^{-1}; density of liquid hydrogen = 0.07 g cm^{-3}; density of liquid oxygen = 1.15 g cm^{-3}; relative atomic masses; H = 1, O = 16)

A10 In about 100 words compare and contrast the combustion of elements and compounds in the laboratory with combustion in rocket engines. Your answer should include equations but neither experimental nor technical detail. Equations and/or structures do not count as words.

(NICCEA)

Answers

1 FORMULAE, EQUATIONS AND AMOUNTS OF SUBSTANCES

Question	Answer	Mark
1	C (Divide 24 by 6.02×10^{23})	1
2	B (From equation, 1 mole of Mg reacts with $500\,cm^3$ of $4\,M$ hydrochloric acid)	1
3	A (Add equations 2 and 3 together $Fe^{2+}(aq) + C_2O_4^{2-}(aq) \rightarrow Fe^{3+}(aq) + 2CO_2(g) + 3e^-$ Multiply this equation by 2 and add this to the first equation. This will remove all of the electrons and give an overall equation $Cr_2O_7^{2-}(aq) + 14H^+(aq) + 2Fe^{2+}(aq) + 2C_2O_4^{2-}(aq) \rightarrow$ $\qquad 2Fe^{3+}(aq) + 4CO_2(g) + 2Cr^{3+}(aq) + 7H_2O(l))$	1

Examiner's tip In question 3, many students will forget that dichromate(VI) oxidises both ethanedioate and iron(II) ions. Getting a correct ionic equation, with the correct stoichiometry, is an important first stage in many calculations.

4	B ($MgSO_4$ and $MgCl_2$: 0.6 moles of chloride combine with 0.3 moles of magnesium, 0.2 moles of sulphate combine with 0.2 moles of magnesium)	1
5 (a)	$3Mn_3O_4 + 8Al \rightarrow 9Mn + 4Al_2O_3$	1
(b)	$4NH_3 + 5O_2 \rightarrow 4NO + 6H_2O$	1
(c)	$Na_2H_2P_2O_7 + 2NaHCO_3 \rightarrow 2Na_2HPO_4 + 2CO_2 + H_2O$	1

Examiner's tip Writing correctly balanced equations is an important attribute of a successful A-level student. Always check any equation you write to make sure it is balanced. An incorrectly balanced equation will cost you a mark!

6 (a)	Mass of oxygen in sample $= 2.32 - 1.68 = 0.64\,g$	
	Number of moles of iron $= 1.68/56 = 0.03$	
	Number of moles of oxygen $= 0.64/16 = 0.04$	1
	Hence, formula of the oxide of iron $= Fe_3O_4$	1

Examiner's tip Many students make a mistake here and write the formula as Fe_4O_3. Remember, 1 mole of an element contains the same number of atoms as 1 mole of another element. There are more oxygen atoms than iron atoms.

(b) (i)	$Fe_2O_3(s) + 3CO(g) \rightarrow 2Fe(l) + 3CO_2(g)$ (One mark is for correct balancing and one for the state symbols)	2

Question	Answer	Mark
(ii)	Reaction with carbon to form (more) carbon monoxide (reaction 3)	1
(iii)	Calcium silicate	1
(iv)	SiO_3^{2-}	1
(c)	Add a suitable reagent, e.g. dilute hydrochloric acid	1
	Iron(III) oxide is basic	1
	and forms a soluble compound (iron(III) chloride)	1

Examiner's tip For a complete answer you might add that the sand is removed by filtration, washed with water and dried. You might also write the equation:

$$Fe_2O_3(s) + 6HCl(aq) \rightarrow 2FeCl_3(aq) + 3H_2O(l)$$

Question	Answer	Mark
7	If the RAM of X is x, using the equation we can say that $(x + 160)\,g$ of XBr_2 would form $(x + 71)\,g$ of XCl_2	
	$$\frac{1.500}{0.89} = \frac{x + 160}{x + 71}$$	
	$x = 58.9$	2
	Using the Periodic Table, the element is cobalt, Co	1

Question	Answer	Mark
8 (a)	The oxidation number of manganese changes from +7 to +2	
	The manganese has been reduced	1
(b)	Number of moles of manganate(VII) used = $28 \times 0.02 \times 10^{-3}$	
	$= 5.6 \times 10^{-4}$ mole	1
	Number of moles of Sn^{2+} in $250\,cm^3$ soln = $5/2 \times 5.6 \times 10^{-4} \times 250/25$	
	$= 0.014$ mole	1
	Mass of tin in the sample used = $0.014 \times 119 = 1.666\,g$	1
	% of tin by mass = $1.666/15 \times 100 = 11.1\%$	1

2 ATOMIC STRUCTURE

Question	Answer	Mark
1	A (Isotopes of the same element with the same mass number would be identical)	1

Examiner's tip Multiple-choice questions including a negative are always more difficult and require special care.

Question	Answer	Mark
2	D (Losing an alpha particle produces $^{207}_{81}$Tl. Losing a beta particle then produces D)	1

Examiner's tip There is a quick way of doing this using the Periodic Table. When an alpha particle is lost the element moves two places to the left and when a beta particle is lost, it moves one place to the right.

3 (a)

species	protons	neutrons	electrons
$^{23}_{11}$Na	11	12	11
$^{16}_{8}$O	8	8	8
$^{18}_{8}$O^{2-}	8	10	10

4

1 mark for all three answers in the top line
1 mark for all three answers in the second line
1 mark for each of the answers in bold in the third line

(b) (i) $1s^2 2s^2 2p^6 3s^1$ — 1

(ii) $1s^2 2s^2 2p^6$ — 1

(c) Any three of the following:
High melting point or high boiling point
Conducts electricity when molten or in aqueous solution
Soluble in water or soluble in polar solvents
Colourless or white — 3

Examiner's tip You are expected to predict physical properties of sodium oxide using your knowledge of other ionic compounds of sodium, e.g. sodium chloride. Make sure you predict physical properties and not chemical ones.

(d) Relative atomic mass = $16 \times 75/100 + 18 \times 25/100$ — 1
   = 16.5 — 1

(e) $44 = C^{16}O_2^+$, $46 = C^{16}O^{18}O^+$, $48 = C^{18}O_2^+$
   All three correct 2 marks, two or one correct 1 mark — 2

4 (a)

	relative mass	relative charge	
proton	1	+1	1
neutron	1	0	1

Examiner's tip The question asks about the main characteristics of particles in the atomic **nucleus**. You get 1 mark for the relative mass and relative charge of each named particle. You must not mention the electron in your answer. It is not in the nucleus.

(b) Same number of protons — 1
   and different number of neutrons — 1

Answers to Unit 2

Question	Answer	Mark

Examiner's tip It is easy to lose a mark here by just writing 'different number of neutrons'. It is a good idea to include an example in your answer, e.g. chlorine-35 contains 17 protons and 18 neutrons and chlorine-37 contains 17 protons and 20 neutrons.

(c) (i) $35 = {}^{35}Cl^+$ $37 = {}^{37}Cl^+$ — 1

$70 = {}^{35}Cl_2^+$ — 1

$72 = ({}^{35}Cl-{}^{37}Cl)^+$ — 1

$74 = {}^{37}Cl_2^+$ — 1

(ii) ${}^{35}Cl$ is more abundant than ${}^{37}Cl$ (In fact three times more) — 1

(d) (i) ${}_0^1 n$ — 1

(ii) ${}_2^4 He$ — 1

(iii) ${}_{-1}^0 e$ — 1

Examiner's tip When balancing nuclear equations such as these, the sum of the subscripts on each side of the equation must be the same. The sum of the superscripts must also be the same on each side of the equation; e.g. in (i), $9 + 4 = 12 + 1$ and $4 + 2 = 6 + 0$.

3 g (1 atom of ${}^4He$ produces 1 atom of ${}^{12}C$, so 4 g produces 12 g, 1 g produces 3 g) — 1

(e) Half-life is the time for the radioactivity level (or number of radioactive nuclei) to reduce to half the original level — 1

Examiner's tip This is frequently misunderstood. For example, in a recent question, the level of radioactivity was said to fall by three-quarters of its original level in 10 hours. The half-life should have been given as 5 hours (10 hours is two half-lives), but the frequent wrong answer was 6.66 hours, i.e. 10 hours was thought to be one and a half half-lives.

5 (a) $4\frac{1}{2}$ hours is three half-lives
$8000 \rightarrow 4000 \rightarrow 2000 \rightarrow \mathbf{1000}$ — 1

(b) No effect on its half-life — 1

Examiner's tip The rate of radioactive decay is dependent only on the number of radioactive nuclei present. Dissolving the sample in a solvent or changing its temperature has no effect. This is a frequent cause of error!

(c) There are many acceptable examples, e.g. the use of gamma radiation from cobalt-60 to kill cancers — 1

6 (a) The ionization energy marked on your graph for K should be less than that for Na — 1

Letts Q&A

Question	Answer	Mark
(b)	Number of shells increases down the group, therefore the outer electron is further from the nucleus	1
	Less energy is required to remove it, therefore ionisation energy decreases	1

Examiner's tip The presence of extra electron shells also increases the shielding of the outer electron.

Question	Answer	Mark
(c)	The pattern of 2 (Na, Mg), 3 (Al, Si, P), 3 (S, Cl, Ar) is noted	1
	The ionisation energy of Group III is lower than Group II because the electron is lost from the *p* shell leaving a full stable *s* orbital. Also the *p* electron in Al is at a higher energy level than the *s* electrons in Na and Mg and is therefore, on average, further from the nucleus.	1

Question	Answer	Mark
7 (a)	RMM = 71 (the highest value on the diagram, corresponding to M^+)	1
(b)	Difference = 19	1
	Therefore atom is F	2
(c)	Difference = 38	1
	Therefore atoms are F_2	1
(d)	Difference = 57	1
	Therefore atoms are F_3	1
(e)	N^+ ion	3
(f)	NF_3	2

Examiner's tip There are many possible sources of error here. If you get a wrong RMM in (a) or make arithmetical errors, you will find getting the correct answers difficult. Marks are available for incorrect but reasonable answers, e.g. CH_7 or OH_3 instead of F.

3 BONDING AND STRUCTURE

Question	Answer	Mark
1	B (Covalent bonds between atoms in the molecule – intramolecular forces – and van der Waals forces between molecules – intermolecular forces	1
2	A	1

Examiner's tip This question is difficult because all of the four statements are correct. It is a matter of deciding which one fits the situation.

Question	Answer	Mark
3	E (The largest difference in electronegativity, 0.7)	1

Question	Answer	Mark

Understanding the idea of electronegativity is very important at A level. You should not try to remember the values. If they are required, they will be given to you in the question, as in question 3 above. You should know, however, that
● the larger the difference in electronegativity, the more polar the bond is;
● electronegativity increases across a period of the Periodic Table;
● electronegativity decreases down a group of the Periodic Table.

---

4   A (90° angles in (i) and (ii), and in (iii) the H–O–H bond angle is approximately 104°, less than tetrahedral angle, due to greater repulsions between non-bonding electron pairs)   **1**

---

5 (a)   Trigonal planar, e.g. BF_3 Tetrahedral, e.g. CH_4
Pyramidal, e.g. NH_3 Bent, e.g. H_2O **4**

If you cannot describe the shape then draw a simple diagram. If you are not sure, try and think of a simple example, e.g. H_2O is an example of a molecule where the central atom has two pairs of bonding electrons and two lone pairs (non-bonding pairs) of electrons.

(b)   Electronegativity is the electron-attracting power of an atom and it is used to predict the extent to which a covalent bond is polar.   **1**

---

6 (a) and (b)

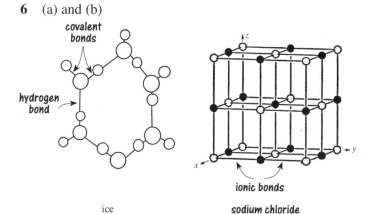

ice ....   sodium chloride ....   iodine ....

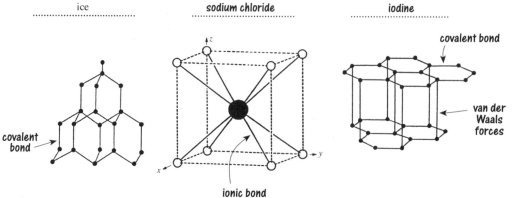

diamond ....   caesium chloride ....   graphite ....

Question	Answer	Mark
	Identify the structures	**4**
	Bonding in ice	**1**
	Bonding in sodium chloride	**1**
	Bonding in iodine	**1**
	Bonding in diamond	**1**
	Bonding in caesium chloride	**1**
	Bonding in graphite	**1**
(c)	Ice, iodine (low melting point)	**1**
	sodium chloride, caesium chloride (high melting point)	**1**
	diamond, graphite (very high melting point)	**1**

Examiner's tip It is impossible to be sure of the exact order, but you should be able to make the distinctions shown above.

Question	Answer	Mark
7 (a)	The mass of molecules increases from H_2S to H_2Te	**1**
	The rise in boiling point from H_2S to H_2Te is due to the increasing number of electrons in the molecules and the consequent rise in van der Waals attractions.	**1**
(b)	Additional hydrogen bonds between water molecules	**1**

Question	Answer	Mark
8 (a) (i)	Pattern of increasing values from left to right	**1**
	Significant jump between second and third values	**1**
(ii)	Any correct statement that attempts to explain the pattern, e.g. successive increase in values for successive removal of electrons closer to the nucleus *or* against increasing pull of ion of increasing positive charge	**1**

Examiner's tip It is not sufficient to write that successive electrons are held more tightly. You are expected to explain the pattern.

Question	Answer	Mark
	Any correct statement that attempts to explain that the jump between the second and third ionization energies is due to the change from the removal of an electron in the outer shell to the removal of an electron from the next shell 'down'	**1**
(b)	*Most likely value*: 0.100 nm	**1**
	Reason: removal of the two outside shell electrons reduces the radius considerably	**1**
(c) (i)	$1s^2 2s^2 2p^6$	**1**
(ii)	One more electron added to complete a shell already almost full makes very little difference to the radius	**1**
(d)	Exchange of electrons to form complete outer shells	**1**
	Calcium atom loses two electrons and fluorine gains one electron	**1**
	Ca^{2+} and F^- forms a compound with a formula CaF_2	**1**

Question	Answer	Mark
9 (a) (i)	See diagram	**1**

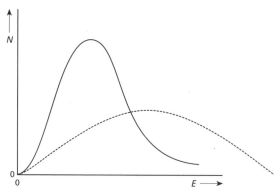

(ii)	By applying pressure of greater than 1.0 atm	**1**
(iii)	Gas (from the position on the graph above)	**1**
(b) (i)	Intermolecular forces (forces of attraction between molecules)	**1**
(ii)	Increasing the temperature increases the average kinetic energies of the molecules	**1**
	More can overcome intermolecular forces and escape	**1**
(c)	See diagram	**1**

> **Examiner's tip**   The dotted line shows the average energies of molecules in a gas at a higher temperature. This is often used in explaining why the rate of a reaction increases with rise in temperature (see Unit 5).

4 ENERGETICS

Question	Answer	Mark
1	C (Energy is released when chlorine atom gains an electron: electron affinity of chlorine is -364 kJ mol^{-1})	**1**

Question	Answer	Mark
2	A (Mass of 1 mole of CH_3OH = 32 g, so 3.2 g is 0.1 moles)	1
3	B (Bonds broken 1 C–H and 1 Br–Br requires energy +414 + 194 = +608 kJ bonds formed 1 C–Br and 1 H–Br liberates energy −280 −366 = −646 kJ Overall enthalpy change = −646 + 608 = −38 kJ mol^{-1})	1

Examiner's tip   This question uses average bond enthalpies. If you took Further tier GCSE Science or Chemistry papers you may have tried questions such as this one.

4 (a)	B	1
(b)	E and C	1

Examiner's tip   Remember that in this kind of grid question, unlike normal multiple-choice questions, there may be more than one correct answer.

5

Examiner's tip   The equation for this reaction is:

$$CH_2\!=\!CH\!-\!CH_2\!-\!CH_2\!-\!CH_2\!-\!CH_3 \; + \; H_2 \; \rightarrow \; CH_3\!-\!CH_2\!-\!CH_2\!-\!CH_2\!-\!CH_2\!-\!CH_3$$
hex-1-ene   hexane

(a) Enthalpy change for the formation of 1 mole of a compound from its elements in their standard states **1**

   Standard state of an element is the state of the element at 298 K (25 °C) and 1 atm **1**

(b)

hex-1-ene + H$_2$ —————— x ——————▶   hexane

−72.4 kJ mol^{-1} −198.6 kJ mol^{-1}

6C (graphite) + 7H$_2$(g) **1**

Using Hess's law −72.4 + x = −198.6 **1**

         x = −126.2 kJ mol^{-1} **1**

Question	Answer	Mark

Examiner's tip   This kind of calculation occurs very often. It is important to draw the
correct triangle. An alternative way of doing the calculation is as follows:
You are trying to find the enthalpy change for

$$C_6H_{12}(g) + H_2(g) \rightarrow C_6H_{14}(g)$$

You are given
1 $6C(s) + 6H_2(g) \rightarrow C_6H_{12}(g)$ $\Delta H = -72.4$ kJ mol^{-1}
2 $6C(s) + 7H_2(g) \rightarrow C_6H_{14}(g)$ $\Delta H = -198.6$ kJ mol^{-1}
You get the required equation by subtracting **1** from **2**. If you do the same with the
ΔH values, you get the correct answer.

(c)   Loss C=C   +612   Gain   C–C   −347
              H–H   +436   2 × C–H   −826
              _____   _____
              +1048   −1173   **1**

$$\Delta H = -125 \text{ kJ mol}^{-1}$$ **1**

(d)   The result obtained from ΔH_f values would be more accurate   **1**
         These are accurate values, whereas average bond energies are only
         average and so values obtained are only approximate   **2**

(e) ΔH_f for CO_2 and H_2O or $\Delta H_{combustion}$ for C and H_2 **2**

---

6 (a)

$$\begin{array}{l}
\text{1st EA + 2nd EA} \nearrow \quad \underline{Ca^{2+}(g) + O^{2-}(g)} \\
\\
\underline{Ca^{2+}(g) + O(g) + 2e^-} \nearrow \\
\\
\qquad\qquad \big\uparrow \text{1st IE + 2nd IE} \qquad\qquad \Big\downarrow \text{lattice energy} \\
\\
\underline{Ca(g) + O(g)} \\
\underline{Ca(g) + \tfrac{1}{2}O_2(g)} \quad \big\uparrow \Delta H_{at} \\
\underline{Ca(s) + \tfrac{1}{2}O_2(g)} \quad \big\uparrow \Delta H_{at} \\
\qquad\qquad \big\downarrow \Delta H_f \qquad\qquad \underline{CaO(s)} \big\downarrow
\end{array}$$

1 mark for state
symbols all correct,
2 marks for all
enthalpy changes correct
(or 1 mark for at least
4 enthalpy changes),
1 mark for all atoms/ions
labelled correctly   **4**

Examiner's tip   Compare the Born–Haber cycle for CaO with the cycle for sodium
chloride in the Revision Summary (page 29). You will notice that the electron affinity
for Cl is negative but for O is positive. Remember the electron affinity for
Cl refers to the equation

$$Cl + e^- \rightarrow Cl^-$$

but the electron affinity for O refers to two processes

$$O + e^- \rightarrow O^-$$
$$O^- + e^- \rightarrow O^{2-}$$

This second step, involving the attraction of a negatively charged electron by a
negatively charged ion, is a very endothermic process.

Question	Answer	Mark
(b)	For CaO: $-635 = +178 + 249 + 1735 + 657 + LE$	1
	$LE = -178 - 249 - 1735 - 657 - 635$	1
	$= -3454$ kJ mol^{-1}	1
	For FeO, by the same method, you should get -3920 kJ mol^{-1}	2
(c)	Fe^{2+} ion is smaller than Ca^{2+}	1
	therefore the attraction is stronger	1

(d)

$FeO + Ca \xrightarrow{\quad x \quad} Fe + CaO$

-278 kJ mol^{-1} -635 kJ mol^{-1}

$Fe + Ca + \frac{1}{2}O_2$

		1
	$-278 + x = -635$	
	$x = -635 + 278$	
	$= -357$ kJ mol^{-1}	1
	This is not a feasible method	1
	as calcium is more expensive than iron	1
(e)	There is no reaction because the rate is too low	1
	and reaction needs to be heated to a higher temperature	1
(f) (i)	The major differences are:	
	3rd ionisation energy	1
	is more endothermic	1
	lattice energy	1
	is more exothermic	1
(ii)	The largest factor is the 3rd ionisation energy	1
(g)	The 3rd ionisation energy	1
	is less endothermic in iron than calcium	1
	because the 3rd electron is not in a complete inner shell	1

7 (a) (i)	Energy exchanged during reaction $= 50.0 \times 3.5 \times 4.17 = 729.8$ J		
	(approximate this to 730 J)		1
(ii)	Amount of each reagent used $= 0.0125$ mol		1
	Standard molar enthalpy change for this reaction $= -729.8 / 0.0125$		
	$= -58380$ J mol^{-1}		
	$= -58.4$ kJ mol^{-1}		1

Examiner's tip   In (a)(i) there is only a single mark and so it is very easy to lose it for a simple error. In this case you will get no marks if you forget the units, J. In (ii) you will lose one of the two marks if your answer is without units or there is no negative sign.

Letts
Q&A

Question			Answer	Mark
(b)	(i)		The ionic reaction is the same in both cases	
			$Ag^+(aq) + Cl^-(aq) \rightarrow AgCl(s)$	1
			For potassium chloride, the same number of chloride ions react as for sodium chloride – hence the same enthalpy change	1
			For calcium chloride, twice the number of chloride ions react as for sodium chloride – hence twice the enthalpy change	2
	(ii)		Possible suggestions include:	
			same volume of calcium chloride solution, but 0.25 mol dm^{-3}; same temperature rise. This would also be true for same volume, and 0.50 mol dm^{-3} concentration	
			or	
			same concentration of calcium chloride solution, but using only 12.5 cm^3; same energy released but as less volume to heat up the temperature rise would be greater. Using same concentration but more than 12.5 cm^3 would give a lower temperature rise; this would still enable the enthalpy change to be calculated provided the volume added was measured.	
			Marking principles would be	
			choice of satisfactory concentration	1
			choice of satisfactory volume	1
			justifying choices in terms of amounts of substances reacting	1
			justifying choices in terms of predicted temperature rise	1

5 KINETICS

Question	Answer	Mark
1	A (Increasing the concentration of the acid will speed up the reaction and possibly increase the final volume of gas produced. A catalyst would not produce more gas, but the same amount faster)	1
2	C (Increasing the pressure of the gaseous reaction mixture by three times increases [C$_2$H$_4$] and [H$_2$] by a factor of three. Using the rate equation $3 \times 3 = 9$)	1
3	A (Remember that reaction rate is inversely proportional to time)	1
4	B (Look up the Arrhenius equation. You will notice it includes activation energy)	1

Question	Answer	Mark
5	D (X in the diagram actually represents the energy advantage of using a catalyst)	1

Examiner's tip This energy profile diagram is very often seen in A-level books and on A-level papers. The diagram shown is for an exothermic reaction because the energy of the reactants is greater than the products; the excess energy is given out to the surroundings. The solid line represents the uncatalysed reaction. You will notice that energy, called the activation energy, has to be provided for the reaction to proceed. If the reaction is carried out at a higher temperature, more of the reactant particles have the necessary energy and the reaction is faster. A catalyst provides an alternative reaction pathway, shown by the dotted line, which requires a lower activation energy. More of the reactant particles possess this energy and so the reaction proceeds faster.

6	D (Compare the 1st and 2nd sets of results: increasing [A] by a factor of 2 while keeping [B] constant increases the rate by a factor of 4. The reaction is second order with respect to [A]. Compare the 1st and 3rd sets of results: increasing [B] by a factor of 2 while keeping [A] constant increases the rate by a factor of 2. The reaction is first order with respect to [B]. It is third order overall)	1

7	(a)	(i) 1	1
		(ii) 1	1
	(b)	Rate = $k[CH_3CO_2C_2H_5][OH^-]$	1

Examiner's tip It is here that the examiner applies consequential marking. The answer given assumes the correct answers in (a). If you gave (a)(i) as 2 and (a)(ii) as 1, you would then have written Rate = $k[CH_3CO_2C_2H_5][OH^-]^2$. Although this is not correct, it does follow consequentially from your previous answers and you would be credited with the mark.

(c) You could use any of the sets of results to get the answer

e.g. $k = \dfrac{8.00 \times 10^{-4}}{(0.064)^2}$ 1

$= 0.195$ 1

Units $\dfrac{\text{mol dm}^{-3}\text{s}^{-1}}{\text{mol}^2 \text{ dm}^{-6}} = \text{mol}^{-1}\text{dm}^3\text{s}^{-1}$ 1

8	(a)	Rate = $k[O_3]^2$	1
		k is the rate constant	1
		the order, 2, is the power to which the molar concentration of O_3 is raised	1

Question	Answer	Mark

(b)　(i)　$mol\,dm^{-3}\,s^{-1}$　　　　　　　　　　　　　　　　　　　　　　　1

(from rate equation, units of rate $=$ units of $k \times$ (units of concentration)^2
$= dm^3\,mol^{-1}\,s^{-1} \times (mol\,dm^{-3})^2)$

　　　(ii)　rate $= 3.38 \times 10^{-5} \times (0.25)^2$　　　　　　　　　　　　　　　1

　　　　　　　$= 0.211 \times 10^{-5}\,mol\,dm^{-3}\,s^{-1}$　　　　　　　　　　　1

Examiner's tip　The units of molar concentration are always $mol\,dm^{-3}$ and the units
of rate are usually $mol\,dm^{-3}\,s^{-1}$. The units of the rate constant will vary according
to the order of the reaction.

(c)　(i)　Mean kinetic energy of molecules increases　　　　　　　　　　1
　　　　　so greater proportion of collisions have energy greater than activation
　　　　　energy　　　　　　　　　　　　　　　　　　　　　　　　　　　　1

　　　(ii)　Catalyst provides a new route　　　　　　　　　　　　　　　　　1
　　　　　of lower activation energy　　　　　　　　　　　　　　　　　　　1

(d)　(i)　Follow the course of the reaction either by titration of hydroxide ion
　　　　　with dilute acid *or* estimation of bromide ion with silver ions　　1
　　　　　maintain $[OH^-]$ constant, vary $[C_2H_5Br]$　　　　　　　　　　1
　　　　　Note *n*-fold change in $[C_2H_5Br]$ leads to an *n*-fold change in rate　1

　　　(ii)

step 1　　$H\ddot{O}^- + C_2H_5 - \overset{\overset{H}{|}}{\underset{\underset{H}{|}}{C}} - Br \xrightarrow{\ slow\ } \left[HO --- C_2H_5 --- Br \right]^-$

step 2　　$\left[HO --- C_2H_5 --- Br \right]^- \xrightarrow{\ fast\ } C_2H_5OH + Br^-$

　　　　　　　　　　　　　　　　　　　　　　　　　　　　　　　　　2

9　(a)　(i)　1　　　　　　　　　　　　　　　　　　　　　　　　　　　　1

　　　　(ii)　rate (or velocity) constant　　　　　　　　　　　　　　　　　1

　　(b)

air-tight
apparatus

gas syringe

flask

reactants

constant temperature
water bath at 40°C

　　　　　　　　　　　　　　　　　　　　　　　　　　　　　　　　　4

(c)　(i)　Draw gradients to the curve at (1) $c = 0.300\,mol\,dm^{-3}$ and
　　　　　(2) $c = 0.600\,mol\,dm^{-3}$

Question	Answer	Mark

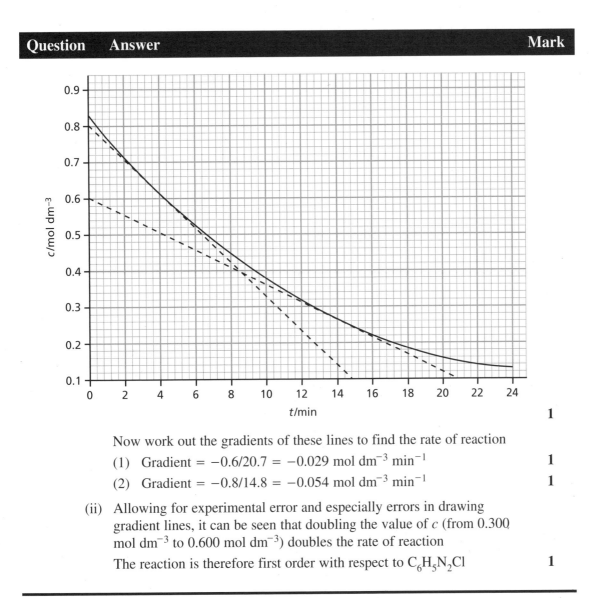

1

Now work out the gradients of these lines to find the rate of reaction

(1) Gradient $= -0.6/20.7 = -0.029$ mol dm^{-3} min^{-1} 1

(2) Gradient $= -0.8/14.8 = -0.054$ mol dm^{-3} min^{-1} 1

(ii) Allowing for experimental error and especially errors in drawing gradient lines, it can be seen that doubling the value of c (from 0.300 mol dm^{-3} to 0.600 mol dm^{-3}) doubles the rate of reaction

The reaction is therefore first order with respect to $C_6H_5N_2Cl$ 1

6 EQUILIBRIUM

Question	Answer	Mark
1	B $\left(K_c = \dfrac{[Y]^2[Z]^3}{[W][X]^2}\right.$ All concentrations are in units of mol dm^{-3})	1
2	E (Ammonia and ammonium chloride will produce an alkaline buffer solution)	1

Examiner's tip   This question highlights one of the frequent errors made by students when answering questions on buffer solutions. You must have a weak acid or alkali and a salt of the weak acid or alkali, e.g. ethanoic acid and sodium ethanoate, or ammonia and ammonium chloride. Many students will ignore the importance of **weak** acid or alkali and give C as the answer.

Question	Answer	Mark
3	D (A larger mass of the organic compound will be extracted by several extractions rather than one large extraction)	1
4	B ($[OH^-] = 10^{-1}$ so $[H^+] = 10^{-13}$)	1
5	A (Increasing the pressure will move the equilibrium to the left)	1

Examiner's tip This question is unusual and will test your understanding. Most of the examples you have met (Haber process, Contact process) involve a decrease in volume and an exothermic reaction in the forward reaction. This example is different and the forward reaction involves an increase in volume and is endothermic in the forward reaction. To get the highest yield of Y and Z you need low pressures and high temperatures.

Question	Answer	Mark
6	A and E (Adding more catalyst does not affect the position of the equilibrium)	1
7 (a)	After 15 seconds [product] = 0.37 mol dm^{-3} Average rate of reaction = 0.37/15 = 0.025 mol dm^{-3} s^{-1}	1
(b)	The system is in dynamic equilibrium	1

Examiner's tip You could work out the rate of loss of reactant in (a). You would get the same answer.

The idea of equilibrium is very difficult to understand.

Remember:

❶ The reaction has not stopped. Both the forward and reverse reactions are proceeding.

❷ The rate of the forward reaction is equal to the rate of the reverse reaction and so the concentrations of the reactants and products do not change.

❸ An equilibrium is unstable and even slight changes in conditions (concentrations of reactants and products, temperature, pressure, etc.) will alter the position of the equilibrium. The equilibrium can move to the right, producing more products, or to the left, producing more of the reactants. Qualitative predictions about an equilibrium can be made using Le Chatelier's principle.

Question	Answer	Mark
8 (a)	Highest yield produced by high pressures and low temperature	2

Examiner's tip You are not expected to know this but to work it out. You will see going down any column that the percentage yield increases with increasing pressure. Going across any row the percentage yield decreases as temperature rises.

Question	Answer	Mark
(b) (i)	According to Le Chatelier's principle, increasing pressure moves the equilibrium to the right. There are four molecules on the left-hand side and two on the right-hand side	1
(ii)	According to Le Chatelier's principle, the yield of ammonia is increased at low temperatures	
	$N_2(g) + 3H_2(g) \rightleftharpoons 2NH_3(g) + heat$	
	Adding heat moves the reaction to the left	1
(c)	Using very high pressures is expensive as costs are involved in compressing the gases and also equipment has to be built to withstand high pressures	1
	Lowering the temperature gives a higher yield of ammonia but the rate of reaction is reduced. Using a catalyst will speed up the reaction	1

Question	Answer	Mark
9 (a) (i)	Increasing the pressure moves the equilibrium to the left	1
	More molecules on the right-hand side than on the left ∴ yield is reduced	1
(ii)	Increasing the temperature moves the equilibrium to the left	1
	Forward reaction is exothermic ∴ yield is reduced	1
(b)	If the temperature is reduced the yield is increased. However, at lower temperatures the rate of reaction will be very slow. There has to be a compromise	1
(c) (i)	To speed up the reaction. (Remember, a catalyst will never produce more, only the same, more quickly!)	1
(ii)	A heterogeneous catalyst provides a surface for the reaction to take place	1
(iii)	A gauze has a very large surface area per unit mass of catalyst	1
(d) (i)	$K_p = \dfrac{p_{N_2O_4}}{p_{NO_2}^{2}}$	1
	Units for pressure are kPa, so units for K_p here are kPa^{-1}	1
(ii)	Partial pressure of N_2O_4, $p_{N_2O_4} = 700 \times 48/100 = 336$ kPa	1
	Partial pressure of NO_2, $p_{NO_2} = 700 - 336 = 364$ kPa	1
	(Remember that the sum of the partial pressures equals the total pressure)	
	Substituting in the formula in (d)(i)	
	$K_p = 336/(364)^2$	
	$= 2.54 \times 10^{-3} kPa^{-1}$	1

Question	Answer	Mark
10 (a) (i)	A weak acid is one in which only a proportion of the molecules are ionised	1
	forming H^+ ions	1

Question	Answer	Mark
(ii)	A neutral solution is one in which both H^+ and OH^- ions are present but in equal concentrations	1 1
(b) (i)	HCl of concentration 0.01 mol dm^{-3} $\quad$ $pH = -\log_{10}[H^+]$ $\qquad\qquad = -\log 0.01$ $\qquad\qquad = 2$	1
(ii)	KOH of concentration 0.02 mol dm^{-3} $\quad [OH^-] = 0.02$ $\qquad\qquad [H^+] = 10^{-14}/0.02$ $\qquad\qquad\qquad = 5 \times 10^{-13}$ $\qquad\qquad pH = -\log_{10}(5 \times 10^{-13})$ $\qquad\qquad\qquad = 12.3$	1
(iii)	Ba(OH)$_2$ of concentration 0.05 mol dm^{-3} $\quad [OH^-] = 2 \times 0.05$ $\qquad\qquad [H^+] = 10^{-14}/2 \times 0.05$ $\qquad\qquad\qquad = 1 \times 10^{-13}$ $\qquad\qquad pH = -\log_{10}(1 \times 10^{-13})$ $\qquad\qquad\qquad = 13$	1
(c)	To make a buffer solution from pure ethanoic acid, add ethanoic acid and sodium hydroxide solution to water. The pH of the buffer solution will depend on the amounts of ethanoic acid and sodium hydroxide added	1
	The sodium hydroxide neutralises part of the acid	1
	$CH_3COOH + NaOH \rightarrow CH_3COO^-Na^+ + H_2O$	
(d)	A buffer solution is one whose pH is relatively resistant to change when small amounts of acid and alkali added to it	1 1
(e)	Any suitable example of a buffer solution in use	1
	An example would be setting up a pH meter. A pH meter is calibrated using a solution of known pH. If the pH of the solution does not change it ensures the pH meter is reading correctly	
11 (a)	A proton donor	1
(b)	*Equation*: $NH_4^+ + OH^- \rightarrow NH_3 + H_2O$	1

> **Examiner's tip** The ionic equation shown is the simplest one which can be written. Ammonium chloride and sodium hydroxide are made of ions: NH_4^+ and Cl^- (from ammonium chloride) and Na^+ and OH^- (from sodium hydroxide). When they are mixed
>
> $$NH_4^+ + Cl^- + Na^+ + OH^- \rightarrow NH_3 + H_2O + Na^+ + Cl^-$$
>
> Ions which appear on both sides, and therefore remain unchanged throughout, may be removed. They are called spectator ions. This gives the simplest ionic equation. If you gave an ionic equation with spectator ions present you would not be penalised.

| | *Explanation*: NH_4^+ is donating H^+ (a proton) to OH^- and is acting as a Brønsted–Lowry acid | 1 |
| | OH^- is accepting a proton from NH_4^+ and is acting as a Brønsted–Lowry base | 1 |

Question	Answer	Mark
(c)	Fully dissociated	1
(d)	$pH = -\log_{10}[H^+]$	1
(e)	0.70	1
(f)	$[OH^-] = 0.05$	1
	$[H^+] = 10^{-14}/[OH^-] = 10^{-14}/0.05$	1
	$= 2 \times 10^{-13}$	1
	$pH = 12.7$	1
(g)	$HF(aq) \rightleftharpoons H^+(aq) + F^-(aq)$	1
	$K_a = \dfrac{[H^+]\,[F^-]}{[HF]}$	1

Examiner's tip There is frequent misunderstanding about the terms *weak acid* and *strong acid*. A strong acid is completely ionised, e.g. $HCl \rightarrow H^+ + Cl^-$. A weak acid is only partly ionised, e.g. HF. HF is very corrosive, attacking glass and causing very severe skin burns.

7 REDOX

Question	Answer	Mark
1	B (O.S. of sulphur changes from +4 to +6)	1
2	A (Chlorine oxidises iodide to iodine. This forms dark blue complex with starch)	1
3	A ($2NO_3^-(aq) + 8H^+(aq) + 3Zn(s) \rightarrow 2NO(g) + 4H_2O(l) + 3Zn^{2+}(aq)$)	1
4	D	1
5	E $\quad Fe(s) \rightleftharpoons Fe^{2+}(aq) + 2e^- \qquad E^\ominus = +0.4\,V$ $\quad Cu^{2+}(aq) + 2e^- \rightleftharpoons Cu(s) \qquad E^\ominus = +0.4\,V$ $\quad Add \qquad E^\ominus = +0.8\,V$	1

Question	Answer	Mark
6 (a) (i)	W is copper, Z is zinc sulphate solution Both required	1
(ii)	An ionic conductor; completes circuit by allowing movement of ions	1
(iii)	$Zn(s) \rightarrow Zn^{2+}(aq) + 2e^-$	1
(iv)	Standard conditions are 298 K and solutions of 1 mol dm^{-3}	1

> **Examiner's tip**   If the cell involves a gas (e.g. hydrogen electrode) the gas should be at 1 atmosphere pressure.

(v)	$+1.10\,V$ $(0.76\,V + 0.34\,V)$	1
(vi)	$Cu^{2+}(aq) + 2e^- \rightarrow Cu(s)$	1
(b) (i)	Dilute acid or alkali, e.g. dilute sulphuric acid or sodium hydroxide	1
(ii)	Suitable apparatus is shown below One mark for apparatus which works and one mark for gases labelled formed at the correct electrodes	2

(c)	$Cu^{2+} + 2e^- \rightarrow Cu$	
	2 faradays of electricity required to deposit 1 mole of copper atoms	
	193 000 coulombs of electricity required to deposit 64 g of copper	1
	193 000/64 × 0.16 coulombs of electricity required to deposit 0.16 g of copper = 482.5 coulombs	1
	Time = 482.5/0.4 s = 1206 s (to the nearest second)	1

7 (a) (i)	$Zn(s) \mid Zn^{2+}(aq) \parallel H^+(aq), H_2(g) \mid Pt$	2

> **Examiner's tip**   You will lose a mark here if you do not include correct state symbols or show the salt bridge ($\parallel$)

(ii)	$+0.76\,V$	1
(iii)	e.m.f. becomes less positive	1
	Le Chatelier's principle predicts the cell reaction is driven to the left-hand side	1

Question	Answer	Mark
(b) (i)	$Zn(s) + 2H^+(aq) \rightarrow Zn^{2+}(aq) + H_2(g)$	**2**

Examiner's tip | Again a mark is lost if state symbols are not given.

(ii)	The hydrogen half-cell	**1**
(iii)	Both oxidation and reduction must occur, so two half-cells required	**2**
(c)	pH increases	**1**
	because H^+ ions removed as $H_2(g)$	**1**
(d)	See diagram II from Unit 7, Question 6(a), page 53	**2**

8 (a) This question uses the equation $\Delta G = \Delta H - T\Delta S$
For the formation of CO, 1 vol $O_2 \rightarrow$ 2 vol CO **1**
Therefore, a net entropy increase due to doubling of gas volume, hence
ΔS is negative **1**
In the case of CO_2, 1 vol $O_2 \rightarrow$ 1 vol CO_2 and ΔS is very small **1**

(b)

Free energy change/kJ mol⁻¹ vs Temperature/K graph showing lines for $C + O_2 \rightarrow CO_2$ and $2C + O_2 \rightarrow 2CO$.

For $T = 0$, $\Delta G = -1115$ kJ mol⁻¹; for $T = 2400$, $\Delta G = -611$ kJ mol⁻¹ **1**
Plot these two points and draw a straight line **1**
Find intercept with $2C + O_2 \rightarrow 2CO$ at between 2260 and 2300 °C **1**

(c) $2/3\Delta G(Al_2O_3)$ at 1300 K $= -842$ kJ mol⁻¹
$\Delta G(CO_2)$ at 1300 K $= -394$ kJ mol⁻¹

Hence $1/2\Delta G(Al_2O_3) = 1/2 \times 3/2 \times -842 = -631.5$ kJ mol⁻¹ **1**
$3/4\Delta G(CO_2) = 3/4 \times -394 = -295.5$ kJ mol⁻¹ **1**
$\Delta G_{reaction} = -295.5 + 631.5 = +336$ kJ mol⁻¹ **2**

8 THE PERIODIC TABLE

Question	Answer	Mark

1	D (Electrolysis of aqueous solutions does not produce an alkali metal deposit at the cathode: $2H^+(aq) + 2e^- \rightarrow H_2(g)$)	1
2	A (All the substances in A are solids and they are hydrolysed to produce acidic solutions)	1
3	D	1
4	B (7 electrons more than the noble gas electronic structure of Ar)	1
5	E (Extra nuclear charge, no extra electron shells)	1
6	E	1
7	B (Mn^{3+} has an electronic structure $[Ar]3d^4$)	1
8	D (Cu(I) has an electronic structure of $1s^2 2s^2 2p^6 3s^2 3p^6 3d^{10}$ Losing a second electron would require breaking down the full $3d$ orbital)	1

Question	Answer	Mark
9	C	1

This question tests some very important chromium chemistry. It mentions two oxidation states of chromium.

Cr^{3+}	chromium(III)	green	oxidation state +3
CrO_4^{2-}	chromate(VI)	yellow	oxidation state +6
$Cr_2O_7^{2-}$	dichromate(VI)	orange	oxidation state +6

Chromate(VI) exists in alkaline solution and dichromate(VI) exists in acidic solution.

$$2CrO_4^{2-}(aq) + 2H^+(aq) \rightleftharpoons Cr_2O_7^{2-}(aq) + H_2O(l)$$

Contrast this with manganate(VI). Manganate(VI) is stable in neutral and alkaline solutions. If heated in acidic solution, manganate(VI) is both oxidised and reduced (disproportionation), becoming Mn(IV) and Mn(VII). Green manganese(VI) ions are converted into MnO_2, a black solid, and purple manganate(VII) solution (MnO_4^-).

10	C (Only these two properties are not expected of Group I metals)	1

---

11 (a)

Na	Mg	Al	Si	P	S	Cl	Ar
NaCl	MgCl$_2$	Al$_2$Cl$_6$ or AlCl$_3$	SiCl$_4$	PCl$_3$	SCl$_2$	Cl$_2$	—
				PCl$_5$	S$_2$Cl$_2$		

$^1/_2$ mark for each correct formula added, rounded up to nearest whole number

4

(b) Details will depend upon the chloride you choose. However, the same marking scheme applies

Reagents used	1
Conditions	1
Method of collecting the product	1

Sample answer Preparation of silicon(IV) oxide

$$Si(s) + 2Cl_2(g) \rightarrow SiCl_4(l)$$

Silicon(IV) chloride is prepared by passing dry chlorine over heated silicon.
Silicon(IV) chloride is collected as a liquid by condensing the vapour using a U tube cooled in cold water.

A lot depends upon the chloride you try to prepare. Aluminium, silicon and phosphorus are probably easier to do.

(c) (i) $NaCl(s) + aq \rightarrow Na^+(aq) + Cl^-(aq)$ 1

(ii) $PCl_3(l) + 3H_2O(l) \rightarrow H_3PO_3(aq) + 3HCl(aq)$ 1

Question	Answer	Mark

(d) (i)

Mg	Cl	O
35.5/24	52.6/35.5	11.9/ 16
1.48	1.48	0.74
2	2	1

1

Empirical formula is Mg_2Cl_2O **1**

(ii) Hydrolysis
produces $MgO \cdot MgCl_2$ **1**

or $2MgCl_2 + H_2O \rightarrow MgCl_2 + MgO + 2HCl$

(e) (i)

1

(ii) Trigonal bipyramid

$$Cl$$

1

12 (a) (i) Lone pair of electrons on oxygen atom **1**

(ii) *M–O bond breaking equation*
$[M(H_2O)_6]^{2+} + NH_3 \rightarrow [M(NH_3)(H_2O)_5]^{2+} + H_2O$ **1**

Reaction type – substitution **1**

O–H bond breaking equation
$[M(H_2O)_6]^{2+} + H_2O \rightarrow [M(OH)(H_2O)_5]^{+} + H_3O^{+}$ **1**

Reaction type – hydrolysis **1**

(b) (i) *Oxidation state* +2 **1**
Coordination number 4 **1**

(ii) $[Pt(NH_3)_2Cl_2] + H_2O \rightarrow [Pt(NH_3)_2Cl(H_2O)]^{+} + Cl^{-}$ **2**

1 mark for correct formula of
the platinum complex and 1 for a
balanced equation

(c) *Explanation of bidentate*   A ligand with two donor atoms **1**

Example of bidentate   Ethanediamine (1,2-diaminoethane)
$H_2NCH_2CH_2NH_2$ **1**

Explanation of multidentate   A ligand with many donor atoms **1**

Example of multidentate   EDTA **1**

Question	Answer	Mark

Examiner's tip   At this stage it is worth giving a little more information about EDTA, which you might use in titrations with metal ions. EDTA (ethenediamine tetraethanoate) is shown below.

$$:^-OOCCH_2 \diagdown \qquad \diagup CH_2COO^-:$$
$$:NCH_2CH_2N:$$
$$:^-OOCCH_2 \diagup \qquad \diagdown CH_2COO^-:$$

It contains six atoms that each have a free lone pair. It is possible for one EDTA molecule to form six coordinate bonds with one metal ion.

13 (a) (i) +2 — 1

(ii) Octahedral — 1

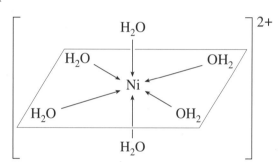

— 1

(b) (i) A buffer solution prevents significant changes in pH — 1

(ii) To ensure all water/ligands have been displaced — 1

(iii) Indicator — 1

(c) There are a number of uses. One is to determine the hardness of water — 1

14 (a)

Halogen	Physical state at room temperature	Colour
Fluorine	gas	yellow with greenish tinge
Chlorine	**gas**	**green/yellow**
Bromine	**liquid**	**red/brown**
Iodine	**solid**	**black**

3

Half a mark for each, rounded up to nearest whole number

(b) (i) $KCl(s) + H_2SO_4(conc) \rightarrow KHSO_4(s) + HCl(g)$ — 1

(ii) $HCl + H_2O \rightleftharpoons H_3O^+ + Cl^-$ — 1
     acid   base   acid   base

Brønsted and Lowry defined an acid as a proton donor and a base as a proton acceptor. Water accepts the proton and is a base. — 1

Question	Answer	Mark
(iii)	Silver nitrate solution contains silver and nitrate ions. When silver nitrate solution is added to a solution of Z (hydrochloric acid, containing chloride ions), a white precipitate of silver chloride is formed	1
	$Ag^+(aq) + Cl^-(aq) \rightarrow AgCl(s)$	1
	The silver chloride dissolves in concentrated ammonia solution forming a soluble complex	
	$AgCl(s) + 2NH_3(aq) \rightarrow [Ag(NH_3)_2]^+(aq) + Cl^-(aq)$	1

Examiner's tip The use of silver nitrate solution to identify the presence of chloride, bromide and iodide is very important. Before testing with silver nitrate solution, the solution being tested should be acidifed. Hydrochloric acid must not be used as this will lead to a positive chloride test every time! The solution should be acidifed with dilute nitric acid.

The following table summarises the results you would expect with chloride, bromide and iodide.

	Colour of precipitate when silver nitrate added	Solubility of precipitate in concentrated ammonia soln
chloride, Cl^-	white	completely soluble
bromide, Br^-	cream	partially soluble
iodide, I^-	yellow	insoluble

Question	Answer	Mark
(c)	$3ClO^-(aq) \rightarrow 2Cl^-(aq) + ClO_3^-(aq)$	1
	O.S. +1 O.S. −1 O.S. +5	1
	Disproportionation occurs when the reactant is both oxidised (chlorate(I) to chlorate(V)) and reduced (chlorate(I) to chloride) in the same reaction	1

Examiner's tip Disproportionation is a very common term in A-level questions.

9 ORGANIC CHEMISTRY

Question	Answer	Mark
1	C (NH_3 with a non-bonding pair of electrons is a good nucleophile)	1
2	C (Nucleophilic addition takes place with aldehydes and ketones and electrophilic addition with alkenes and alkynes)	1
3	C (Addition of hydrogen bromide to propene will use Markownikoff's rule – more negative part (Br) will add on to carbon atom attached to least number of hydrogen atoms)	1

$$\begin{array}{l} CH_3CHCH_3 \\ \quad | \\ \quad Br \end{array} \quad \text{2-bromopropane} \qquad CH_3CH_2CH_2Br \quad \text{1-bromopropane}$$

Question	Answer	Mark
4	D (All hydrogen atoms in propanone are in equivalent positions. In other compounds there are several different positions for hydrogen atoms, which would give two or more peaks)	1

$$H-\overset{\overset{\displaystyle H}{|}}{\underset{\underset{\displaystyle H}{|}}{C}}-\overset{\overset{\displaystyle H}{|}}{\underset{\underset{\displaystyle H}{|}}{C}}-\overset{\overset{\displaystyle H}{|}}{\underset{\underset{\displaystyle H}{|}}{C}}-O-H$$

A

$$H-\overset{\overset{\displaystyle H}{|}}{\underset{\underset{\displaystyle H}{|}}{C}}-\overset{\overset{\displaystyle H}{|}}{\underset{\underset{\underset{\displaystyle H}{|}}{\displaystyle O}}{C}}-\overset{\overset{\displaystyle H}{|}}{\underset{\underset{\displaystyle H}{|}}{C}}-H$$

B

$$H-\overset{\overset{\displaystyle H}{|}}{\underset{\underset{\displaystyle H}{|}}{C}}-\overset{\overset{\displaystyle H}{|}}{\underset{\underset{\displaystyle H}{|}}{C}}-C\overset{\displaystyle H}{\underset{\displaystyle O}{\diagdown}}$$

C

$$H-\overset{\overset{\displaystyle H}{|}}{\underset{\underset{\displaystyle H}{|}}{C}}-\overset{\overset{\displaystyle O}{\|}}{C}-\overset{\overset{\displaystyle H}{|}}{\underset{\underset{\displaystyle H}{|}}{C}}-H$$

D

5	(a)	D	1
	(b)	A (Propanal oxidised to propanoic acid)	1
	(c)	C (Methyl ethanoate is an ester which is hydrolysed with warm sodium hydroxide solution, forming methanol and ethanoic acid)	1

6	(a)	B	1
	(b)	D (Addition of H_2O)	1

7		E	1

8		E (Geometric isomerism caused by lack of rotation at $C=C$)	1

9 (a)

$$H-\overset{\overset{\displaystyle H}{|}}{\underset{\underset{\displaystyle H}{|}}{C}}-\overset{\overset{\displaystyle Br}{|}}{\underset{\underset{\displaystyle H}{|}}{C}}-Br$$

1,1-dibromoethane is a structural isomer of 1,2-dibromoethane

1

(b)

$$\overset{CH_3}{\underset{H}{\diagup}}C=C\overset{H}{\underset{CH_3}{\diagdown}}$$ *trans* form

$$\overset{H}{\underset{CH_3}{\diagup}}C=C\overset{H}{\underset{CH_3}{\diagdown}}$$ *cis* form

1

(c)

$$CH_3\overset{\overset{\displaystyle H}{|}}{\underset{\underset{\displaystyle C_3H_7}{}}{\overset{\displaystyle C}{\diagup}}}\overset{}{\diagdown}C_2H_5$$

1

Examiner's tip   The central carbon atom, if it is to be chiral, must be attached to four different groups. The simplest groups which can be attached are $H-$, CH_3-, C_2H_5- and C_3H_7- .

Question	Answer	Mark
10 (a)	**A** 1-Bromo-2-methylbenzene	1
	B (Bromomethyl)benzene	1

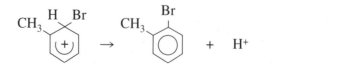
(b) **A** React methylbenzene with bromine

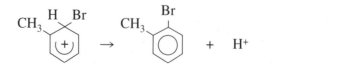

Heat the mixture out of contact with light and in the presence of a suitable halogen carrier (catalyst) such as aluminium bromide, $AlBr_3$, or iron(III) bromide, $FeBr_3$ 1

Electrophilic substitution reaction takes place   1

The halogen carrier causes a polarization of the halogen molecule

$$\overset{\delta+}{Br} \text{-----} Br \text{-----} \overset{\delta-}{AlBr_3}$$

and the positive end of the complex attacks the benzene ring and forms a complex with the electrons of the benzene ring   1

This complex breaks down to give the product

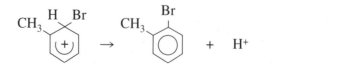

1

Question	Answer	Mark

B React methylbenzene with bromine

$$CH_3 + Br_2 \rightarrow CH_2Br + HBr$$

in the presence of UV light (sunlight) | 1

Free radical substitution reaction takes place | 1

$Br-Br \rightarrow 2Br\bullet$ bromine free radicals | 1

$C_6H_5CH_3 + Br\bullet \rightarrow C_6H_5CH_2\bullet + HBr$

$C_6H_5CH_2\bullet + Br-Br \rightarrow C_6H_5CH_2Br + Br\bullet$ | 1

(c)

Examiner's tip At this stage it would be useful to refer to the **Examiner's tip** on page 108 about silver halides. The answer relies on you knowing that the bromine atom in **A** is difficult to remove but the bromine atom in **B** is easily removed.

Heat each compound with an aqueous solution of sodium hydroxide | 1

Acidify the resulting solutions with dilute nitric acid | 1

Add silver nitrate solution to each solution | 1

Cream precipitate of silver bromide formed with **B** but not with **A** | 1

(d) Reflux with aqueous sodium hydroxide solution, forming phenylmethanol | 1

$$CH_2Br + OH^- \rightarrow CH_2OH + Br^-$$

| 2

Oxidise the alcohol with acidified potassium dichromate(VI) | 1

$$CH_2OH \xrightarrow{+ [O]} CHO \xrightarrow{+ [O]} COOH$$

| 2

---

11 (a)

$$A \quad H-\underset{\underset{H}{|}}{\overset{\overset{H}{|}}{C}}-\underset{\underset{H}{|}}{\overset{\overset{H}{|}}{C}}-\underset{\underset{H}{|}}{\overset{\overset{H}{|}}{C}}-\underset{\underset{H}{|}}{\overset{\overset{H}{|}}{C}}-OH \quad or \quad H-\underset{\underset{H}{|}}{\overset{\overset{H}{|}}{C}}-\underset{\underset{H}{|}}{\overset{\overset{H}{|}}{C}}-\underset{\underset{H}{|}}{\overset{\overset{OH}{|}}{C}}-\underset{\underset{H}{|}}{\overset{\overset{H}{|}}{C}}-H$$

B

$$H-\overset{\overset{H}{|}}{\underset{\underset{H}{|}}{C}}-\overset{\overset{H-\overset{|}{\underset{|}{C}}-H}{|}}{\underset{\underset{OH}{|}}{C}}-\overset{\overset{H}{|}}{\underset{\underset{H}{|}}{C}}-H$$

| 2

A is oxidised by acidified potassium dichromate(VI), and so is a primary or secondary alcohol. **B** is a tertiary alcohol

Question	Answer	Mark

(b)

C   H−C−C−C−C (with H's on carbons and =O, −H group)   D   H−C−C−C−C−H (with H's and =O)

2

C and **D** are carbonyl compounds (aldehydes or ketones). **C** reduces ammonical silver nitrate solution (Tollens reagent) and so is an aldehyde. **D** does not and is therefore a ketone

(c)

E   H−C−C−C (with H's, =O and OH group)

propanoic acid

F   H−C−C (with H's, =O and OCH₃)   or   H−C−O−C−C (with H's, =O and H)   or   H−C (with =O and OCH₂CH₃)

methyl ethanoate   methoxyethanal   ethyl methanoate

2

Carboxylic acids and esters are isomeric. **E** reacts with sodium hydrogencarbonate and is therefore the acid

(d)

G   (benzene ring with CH₃ and NH₂ groups)   H   (benzene ring with CH₂NH₂ group)

2

The amine group must be attached directly to the benzene ring in **G** so that a diazonium salt is produced when treated with nitrous acid below 5°C

(e)

I   H−C−C≡C−C−H (with H's)

but-2-yne

J   (cyclobutene structure with C=C)   or   (1,3-butadiene structure with two C=C)

cyclobutene   1,3-butadiene

2

Question	Answer	Mark
12 (a)	A: O–H (or C–O)	1
	B: C–H	1

Look at the differences between the two spectra and between the two molecules.

$$
\begin{array}{cc}
\text{H} & \text{H} \\
| & | \\
\text{H}-\text{C}-\text{C}-\text{H} \\
| & | \\
\text{H} & \text{H}
\end{array}
\qquad
\begin{array}{cc}
\text{H} & \text{H} \\
| & | \\
\text{H}-\text{C}-\text{C}-\text{O}-\text{H} \\
| & | \\
\text{H} & \text{H}
\end{array}
$$

ethane   ethanol

The differences in the spectra must be due to O–H and C–O bonds.

(b)	Ethanol + potassium bromide + concentrated sulphuric acid	1
	forms bromoethane	1
	bromoethane + ammonia (dissolved in ethanol) and reflux	1
	Alternative answer	
	The first two marks can be obtained by a different route:	
	Dehydration of ethanol using concentrated sulphuric acid to produce ethene	1
	Electrophilic addition of hydrogen bromide to ethene produces bromoethane	1
	The third mark is as before	
(c)	Amine group	1
	N–H bond similar to O–H bond (similar polarity) *or* broadening of peak in spectra due to hydrogen bonding present in both groups	1

13 (a) (i)	Ethanoyl chloride and aluminium chloride	1

There are four examples of electrophilic substitution reactions you meet commonly on examination papers: **nitration**, **sulphonation**, **halogenation** and **Friedel–Crafts** reactions. Friedel–Crafts reactions can be carried out with haloalkanes to produce alkylbenzenes, e.g. methylbenzene,

$$
\bighexagon + CH_3Cl \xrightarrow{AlCl_3} \bighexagon\!\!\!\!\!-CH_3 + HCl
$$

or acyl chlorides to produce ketones, e.g. phenylethanone.

(ii)	Dry or anhydrous	1
(iii)	$C_6H_6 + CH_3COCl \rightarrow C_6H_5COCH_3 + HCl$	1
(iv)	$CH_3\overset{+}{C}{=}O$	1

(b) (i)

$$
CH_3COC_6H_5 + HCN \rightarrow CH_3-\underset{\underset{CN}{|}}{\overset{\overset{OH}{|}}{C}}-C_6H_5
$$

1

Question	Answer	Mark
(ii)	$$R^2 \diagdown \atop R^1 \diagup C = O \quad :CN^- \longrightarrow \quad R^2 \diagdown \atop R^1 \diagup C \diagup CN \atop \diagdown O^- \quad \xrightarrow{H^+} \quad R^2 \diagdown \atop R^1 \diagup C \diagup CN \atop \diagdown OH$$	2
(iii)	$$R^1 \diagdown \atop C \diagdown OH \atop R^2 | CN \qquad NC \diagdown \atop HO \diagup C \diagdown R^1 \atop R^2$$	1
	Equal amounts of *d*- and *l*-forms produces a racemic mixture, so no overall optical activity	1
(c) (i)	$$H-\underset{H}{\overset{H}{C}}-\underset{H}{\overset{H}{C}}-\underset{H}{\overset{H}{C}}-\overset{O}{\underset{\parallel}{C}}-\underset{H}{\overset{H}{C}}-H$$	1
(ii)	$CH_3-CH_2-CH_2-C-CH_3$ ... (two hydrazone structures with 2,4-dinitrophenyl groups)	2

10 SOME NATURAL AND SYNTHETIC MATERIALS

Question	Answer	Mark
1	B (Proteins are made up from different amino acids)	1
2	C (A condensation reaction occurs each time molecules join together in forming the polymer)	1
3	D	1

Question			Answer	Mark
4	(a)	(i)	Two sugar molecules joined together	1
		(ii)	Broken down using water	1
	(b)		glucose / fructose structures	2
	(c)	(i)	There will be equal concentrations of the two sugars	1
			Sucrose is made up of 1 molecule of glucose and 1 molecule of fructose	1
		(ii)	Presence of chiral centre(s)	1
		(iii)	(+), clockwise rotation of plane-polarized light as viewed down the polarimeter; (−), opposite	2

Question			Answer	Mark
5	(a)	(i)	Suitable catalyst, e.g. titanium(IV) chloride, $TiCl_4$, with $Al(CH_2CH_3)_2Cl$ or triethylaluminium	1

> **Examiner's tip** Ziegler catalysts give an alternative method of polymerising alkenes to produce addition polymers. Before the development of these catalysts, polymerisation was done at high temperature and pressure. Ziegler catalysts work at low temperatures in a suitable organic solvent.

		(ii)	A polymer with a regular structure	1
	(b)	(i)	isotactic structure diagram	2

> **Examiner's tip** Your diagram must indicate a three-dimensional structure. If it does not you will only get 1 mark. An example of a diagram getting only 1 mark is
>
> $$-\overset{\overset{\displaystyle H}{|}}{\underset{\underset{\displaystyle CH_3}{|}}{C}}-\overset{\overset{\displaystyle H}{|}}{\underset{\underset{\displaystyle CH_3}{|}}{C}}-\overset{\overset{\displaystyle H}{|}}{\underset{\underset{\displaystyle CH_3}{|}}{C}}-\overset{\overset{\displaystyle H}{|}}{\underset{\underset{\displaystyle CH_3}{|}}{C}}-$$

		(ii)	Methyl groups are randomly orientated *or* not all on the same side of the polymer chain	1
		(iii)	Isotactic has regular structure therefore is tough and crystalline	1
			Atactic is irregular so is soft and rubbery	1
		(iv)	An example is carpet fibre, which needs to be tough	1

11 COMPREHENSION QUESTION

Question	Answer	Mark
A1 (a)	A free radical is an atom or group of atoms containing an unpaired electron. They are very reactive and can be produced by homolytic fission of a covalent bond. e.g. $H—H \rightarrow H\bullet + H\bullet$ hydrogen free radicals	2
(b)	A helium atom contains two electrons. A helium atom in its ground state has both electrons in the lowest energy level – the $1s$ level.	2
(c)	A catalyst alters the rate of a chemical reaction. A burn-rate catalyst controls the rate of combustion of the fuel.	2

Examiner's tip You must look back in the passage and answer the question in context. There are 2 marks for each part – 1 for a brief answer or 2 for a detailed answer related to the passage.

A2 (a)	The lilac or purple colour is due to potassium nitrate.	1
(b)	To prove the presence of carbonate in the residue, add dilute hydrochloric acid to the residue. Effervescence (bubbling) should be seen and the colourless gas produced (carbon dioxide) should turn limewater milky. $K_2CO_3 + 2HCl \rightarrow 2KCl + CO_2 + H_2O$ $Ca(OH)_2 + CO_2 \rightarrow CaCO_3 + H_2O$	3

Examiner's tip Remember the lilac-coloured flame produced by any potassium compound in a flame test. The two equations will not receive extra marks but, if you can write them correctly, they create the right impression.

A3 (a)	symmetrical $\underset{H \quad\quad\ H}{\overset{CH_3 \quad\quad\ CH_3}{N=N}}$ unsymmetrical $\underset{CH_3 \quad\quad\ H}{\overset{CH_3 \quad\quad\ H}{N=N}}$	3
(b)	1,1-dimethylhydrazine	2

A4	$H_2 + F_2 \rightarrow 2HF$ This could be considered as: $F_2 + 2e^- \rightarrow 2F^-$ Fluorine gains electrons – reduction $H_2 \rightarrow 2H^+ + 2e^-$ Hydrogen loses electrons – oxidation The reaction involves oxidation but does not involve oxygen.	2

Examiner's tip It is a frequent mistake for students to remember only the simple definition of oxidation learnt in the lower school, namely that oxidation is the addition of oxygen. Oxidation, when defined more correctly as a process involving loss of electrons, opens up many more possibilities. Remember: **leo – l**oss of **e**lectrons is **o**xidation.

Question	Answer	Mark
A5 (a)	Ammonium perchlorate contains NH_4^+ and ClO_4^- ions. The formula is NH_4ClO_4	2
(b)	Exhaust gases from the rocket may contain acidic gases such as hydrogen chloride and possibly oxides of chlorine which will be highly acidic.	2

A6 A helium atom has an electron configuration of $1s^2$ in its ground state.

$$3s \underline{\qquad\qquad}$$
$$2p \underline{\qquad\qquad}$$
$$2s \underline{\qquad\qquad}$$
$$1s \underline{\quad}\overline{\underline{\downarrow\uparrow}}\underline{\quad}$$

If the atom is excited, one or more electrons can be promoted to a higher energy level.

$$3s \underline{\qquad\qquad}$$
$$2p \underline{\qquad\qquad}$$
$$2s \underline{\quad}\overline{\underline{\uparrow}}\underline{\quad}$$
$$1s \underline{\quad}\overline{\underline{\uparrow}}\underline{\quad}$$

2

A7 Like other metals, metallic hydrogen would be a very good conductor of electricity. Hydrogen ions (H^+ or protons) will be closely packed and free electrons will be able to move through the solid.

3

A8 $2C_{12}H_{26} + 37O_2 \rightarrow 24CO_2 + 26H_2O$

2

Examiner's tip This is a very difficult equation to write and balance correctly. The products are carbon dioxide and water, because the question refers to complete combustion. There is 1 mark for getting the four species in the right place and 1 mark for balancing it correctly. You would not be expected to remember this equation and you certainly would not guess it! You have to work it out.

Starting with $\quad C_{12}H_{26} + O_2 \rightarrow CO_2 + H_2O$

try 12 carbon dioxide molecules to give 12 carbon atoms to match the left-hand side, and then 13 water molecules to match the 26 hydrogen atoms on the left-hand side.

$$C_{12}H_{26} + O_2 \rightarrow 12CO_2 + 13H_2O$$

Now add up the number of oxygen atoms on each side. There are $(12 \times 2) + 13 = 37$ on the right-hand side. To get the same number on the left-hand side you would need to have $18^1/_2$ molecules of oxygen. We double all the way through to remove any fractions. Now go back and try this again. This is important. You will be doing similar things elsewhere in the course.

Question	Answer	Mark

A9   The combustion of hydrogen is summarised by the equation:

$$2H_2(g) + O_2(g) \rightarrow 2H_2O(g)$$

Mass of hydrogen = volume (in cm^3) × density (in $g\ cm^{-3}$)
$$= 253\,000\,000 \times 0.07 = 17\,710\,000\,g$$

Number of moles of hydrogen = mass/mass of 1 mole of H_2
$$= 17\,710\,000/2 = 8\,855\,000 \text{ moles}$$

Assuming there is only steam in the exhaust, there can be no excess hydrogen or oxygen. From the equation, the number of moles of steam produced is the same as the number of moles of hydrogen used, i.e. $8\,855\,000$ moles

Use the formula given

$$pV = nRT$$
$$V = 8\,855\,000 \times 0.082 \times 3280$$
$$= 2.38 \times 10^9\ dm^3$$

4

Examiner's tip This is a difficult calculation. You do not need to use the density of oxygen. In fact, if you try you may run into difficulties! Remember the temperature must be in kelvin and watch the units for R.

A10   A sample answer

The combustion of elements and compounds in the laboratory tends to produce similar products to that in a rocket engine. However, due to lower temperatures and the fact that, in the laboratory, oxygen is mixed with other unreactive gases such as nitrogen, the reactions are much slower.
    Burning hydrocarbons in excess oxygen produces carbon dioxide and water. In a limited amount of oxygen carbon monoxide and water are produced.

$$CH_4 + 2O_2 \rightarrow CO_2 + 2H_2O$$
$$2CH_4 + 3O_2 \rightarrow 2CO + 4H_2O$$

    Nitrogen hydrides (ammonia and hydrazine) do not burn in air but burn in oxygen. Nitrogen is produced.

$$4NH_3 + 3O_2 \rightarrow 2N_2 + 6H_2O$$
$$N_2H_4 + O_2 \rightarrow N_2 + 2H_2O$$

These hydrides, which burn in a rocket engine, do not normally burn in the laboratory.

8

101 words

Examiner's tip This exercise requires you to answer in a brief form selecting from all the chemistry you know. Try to make a couple of points and give a couple of examples to support your points. You do not have to be as close to 100 words as this, between 90 and 110 would be acceptable.